Molecular Magnetism of Lanthanides Complexes and Networks

Special Issue Editor

Kevin Bernot

MDPI • Basel • Beijing • Wuhan • Barcelona • Belgrade

MDPI

Special Issue Editor
Kevin Bernot
Institut des Sciences Chimiques de Rennes (ISCR)
France

Editorial Office
MDPI
St. Alban-Anlage 66
Basel, Switzerland

This edition is a reprint of the Special Issue published online in the open access journal *Magnetochemistry* (ISSN 2312-7481) from 2016–2017 (available at: http://www.mdpi.com/journal/magnetochemistry/special_issues/lanthanides_complexes).

For citation purposes, cite each article independently as indicated on the article page online and as indicated below:

Lastname, F.M.; Lastname, F.M. Article title. *Journal Name* **Year**, *Article number*, page range.

First Editon 2018

ISBN 978-3-03842-987-6 (Pbk)
ISBN 978-3-03842-988-3 (PDF)

Table of Contents

About the Special Issue Editor

Kevin Bernot is an Associate Professor at the National Institute for Applied Sciences (INSA) in Rennes, France. His research team is part of the Institut des Sciences Chimiques de Rennes (ISCR) that comprises 400 chemists covering all fields of chemistry.

His research focuses on Lanthanides Coordination Chemistry as a tool for the synthesis of magnetic and luminescent material. The targeted applications are Single-Molecule Magnets (SMM), Single-Chain Magnets (SCM), hybrid radical-lanthanides complexes, luminescent complexes and magneto-luminescent correlations. Dr. Bernot has published over 75 papers (WoS H = 29, citations > 3000).

Dr. Bernot was awarded the best PhD European thesis in Molecular Magnetism (2008), the PEDR French fellowship for excellence in research (2015), and, since 2017, he is a Junior Member of Institut Universitaire de France (IUF), which is a five-year excellence membership awarded each year to 70 assistant professors across all research fields.

magnetochemistry

MDPI

Editorial

Molecular Magnetism of Lanthanides Complexes and Networks

Kevin Bernot

Institut des Sciences Chimiques de Rennes (ISCR), 20 av. des Buttes de Coesmes, 35708 Rennes, France; kevin.bernot@insa-rennes.fr; Tel.: +33-2-23-23-84-34

Received: 24 July 2017; Accepted: 30 July 2017; Published: 2 August 2017

Lanthanides ions allows for the design of remarkable magnetic compounds with unique magnetic properties. One of their assets is that they can give rise easily to multi-functional materials. Such multi-functionality is found in the collection of papers of this Special Issue with contributions that highlight the unprecedented magnetic properties of lanthanide-based molecules together with chirality [1], luminescence [2,3] or electrical conductivity properties [4]. Innovative synthetic routes such as the use of helical [1,5] or protonated ligands [6] together with cutting edge characterization techniques of 4f-SMM are presented [7].

First of all, this Issue features a remarkable review article from Oliver Waldman and co-workers [7] that highlights the power of neutron scattering studies for the understanding of the magnetic behavior of 4f-SMMs. This extended and remarkably accurate work provides a deep and comprehensive perspective on this technique by analyzing, among others, one of the most famous molecules in our field, the $Tb_2(\mu-N_2{}^{3-})$ dimer.

Then, a study of Jerôme Long and co-workers [3] details how the analysis of luminescent properties of Dy-SMMs can be useful for understanding their magnetic properties. The authors use the very simple molecule $[Dy(NO_3)_3(H_2O)_4]\cdot2H_2O$ which, though often used as a precursor in the design of Dy-SMM, has never been deeply characterized.

Miki Hasegawa, Takashi Kajiwara and co-workers [5] present a nice 4f-SMM family based on a helical ligand in which, surprisingly, not only the Dy^{III} but also the Nd^{III} derivative show SMM behavior [5]. Helicity is also the topic of the work of Boris Le Guennic and co-workers [1] in which SMM behavior is observed on a racemic form of a helicene-based molecule with a remarkable magnetic hysteresis opening.

Albert Escuer, Spyros Perlepes and co-workers [6] report a new approach to the widely used triethanolamine ligand that gives rise to a family of Ln^{III} complexes in which the Dy^{III} derivative behave as a SMM.

Pierre Rabu, Emilie Delahaye and co-workers [2] show how synthetic conditions can influence the creation of magnetic hybrid networks in which the Sm^{III} and Pr^{III} adducts depict luminescent properties.

Masahiro Yamashita and co-workers [4] present a very appealing hybrid material in which partially oxidized BEDT-TTF molecules crystallize together with Dy^{III} precursors to form a compound in which both SMM behavior and electrical conductivity can be observed.

I hope that this Special Issue will be pleasant and useful to the readers of *Magnetochemistry* and I wish this new open access journal all the best for its future.

I am thankful to the *Magnetochemistry* Editor, Carlos J. Gomez-Garcia, for his confidence in giving me the opportunity to guest edit this Special Issue. I am also thankful to the MDPI editorial team for their professionalism and reactivity. I also want to acknowledge the work of all referees that accepted to spend their time to judge, comment and finally enhance the quality of the papers. Finally, and most of all, I would like to thank the authors for their valuable contributions to this Issue.

Conflicts of Interest: The authors declare no conflict of interest.

References

1. Fernandez-Garcia, G.; Flores Gonzalez, J.; Ou-Yang, J.-K.; Saleh, N.; Pointillart, F.; Cador, O.; Guizouarn, T.; Totti, F.; Ouahab, L.; Crassous, J.; et al. Slow Magnetic Relaxation in Chiral Helicene-Based Coordination Complex of Dysprosium. *Magnetochemistry* **2017**, *3*, 2. [CrossRef]
2. Farger, P.; Leuvrey, C.; Gallart, M.; Gilliot, P.; Rogez, G.; Rabu, P.; Delahaye, E. Elaboration of Luminescent and Magnetic Hybrid Networks Based on Lanthanide Ions and Imidazolium Dicarboxylate Salts: Influence of the Synthesis Conditions. *Magnetochemistry* **2017**, *3*, 1. [CrossRef]
3. Mamontova, E.; Long, J.; Ferreira, R.; Botas, A.; Luneau, D.; Guari, Y.; Carlos, L.; Larionova, J. Magneto-Luminescence Correlation in the Textbook Dysprosium(III) Nitrate Single-Ion Magnet. *Magnetochemistry* **2016**, *2*, 41. [CrossRef]
4. Shen, Y.; Cosquer, G.; Breedlove, B.; Yamashita, M. Hybrid Molecular Compound Exhibiting Slow Magnetic Relaxation and Electrical Conductivity. *Magnetochemistry* **2016**, *2*, 44. [CrossRef]
5. Wada, H.; Ooka, S.; Iwasawa, D.; Hasegawa, M.; Kajiwara, T. Slow Magnetic Relaxation of Lanthanide(III) Complexes with a Helical Ligand. *Magnetochemistry* **2016**, *2*, 43. [CrossRef]
6. Mylonas-Margaritis, I.; Mayans, J.; Sakellakou, S.-M.; P. Raptopoulou, C.; Psycharis, V.; Escuer, A.; P. Perlepes, S. Using the Singly Deprotonated Triethanolamine to Prepare Dinuclear Lanthanide(III) Complexes: Synthesis, Structural Characterization and Magnetic Studies. *Magnetochemistry* **2017**, *3*, 5. [CrossRef]
7. Prša, K.; Nehrkorn, J.; Corbey, J.; Evans, W.; Demir, S.; Long, J.; Guidi, T.; Waldmann, O. Perspectives on Neutron Scattering in Lanthanide-Based Single-Molecule Magnets and a Case Study of the $Tb_2(\mu-N_2)$ System. *Magnetochemistry* **2016**, *2*, 45.

magnetochemistry

MDPI

Review

Perspectives on Neutron Scattering in Lanthanide-Based Single-Molecule Magnets and a Case Study of the Tb$_2$(μ-N$_2$) System

Krunoslav Prša [1], Joscha Nehrkorn [1,2], Jordan F. Corbey [3], William J. Evans [3], Selvan Demir [4,5], Jeffrey R. Long [4,6,7], Tatiana Guidi [8] and Oliver Waldmann [1,*]

[1] Physikalisches Institut, Universität Freiburg, D-79104 Freiburg, Germany; krunoslav.prsa@physik.uni-freiburg.de (K.P.); nehrkorn@uw.edu (J.N.)
[2] Department of Chemistry, University of Washington, Seattle, WA 98195, USA
[3] Department of Chemistry, University of California, Irvine, CA 92617, USA; jcorbey@uci.edu (J.F.C.); wevans@uci.edu (W.J.E.)
[4] Department of Chemistry, University of California, Berkeley, CA 94720, USA; selvan.demir@chemie.uni-goettingen.de (S.D.); jrlong@berkeley.edu (J.R.L.)
[5] Institut für Anorganische Chemie, Georg-August-Universität Göttingen, Tammannstraße 4, 37077 Göttingen, Germany
[6] Department of Chemical and Biomolecular Engineering, University of California, Berkeley, CA 94720, USA
[7] Materials Sciences Division, Lawrence Berkeley National Laboratory, Berkeley, CA 94720, USA
[8] ISIS Facility, Rutherford Appleton Laboratory, Chilton, Didcot, Oxfordshire OX11 0QX, UK; tatiana.guidi@stfc.ac.uk
[*] Correspondence: oliver.waldmann@physik.uni-freiburg.de; Tel.: +49-761-203-5717

Academic Editor: Kevin Bernot
Received: 8 November 2016; Accepted: 25 November 2016; Published: 14 December 2016

Abstract: Single-molecule magnets (SMMs) based on lanthanide ions display the largest known blocking temperatures and are the best candidates for molecular magnetic devices. Understanding their physical properties is a paramount task for the further development of the field. In particular, for the poly-nuclear variety of lanthanide SMMs, a proper understanding of the magnetic exchange interaction is crucial. We discuss the strengths and weaknesses of the neutron scattering technique in the study of these materials and particularly for the determination of exchange. We illustrate these points by presenting the results of a comprehensive inelastic neutron scattering study aimed at a radical-bridged diterbium(III) cluster, Tb$_2$(μ-N$_2$$^{3-}$), which exhibits the largest blocking temperature for a poly-nuclear SMM. Results on the Y^{III} analogue Y$_2$(μ-N$_2$$^{3-}$) and the parent compound Tb$_2$(μ-N$_2$$^{2-}$) (showing no SMM features) are also reported. The results on the parent compound include the first direct determination of the lanthanide-lanthanide exchange interaction in a molecular cluster based on inelastic neutron scattering. In the SMM compound, the resulting physical picture remains incomplete due to the difficulties inherent to the problem.

Keywords: single-molecule magnet; lanthanide ions; inelastic neutron scattering; ligand field; Ising model; magnetic exchange

1. Introduction

Single-molecule magnets (SMMs) based on lanthanide ions offer an exciting development towards their potential practical usage, and this field has accordingly attracted enormous attention recently. In particular, large magnetic moments linked with the 4f electronic shell and large anisotropy enable higher blocking temperatures (T_B) than those previously achieved in SMMs containing transition-metal ions [1–3]. Lanthanide-containing molecules are also promising candidates in many other areas, ranging from magneto-calorics, over exotic quantum many-body states to quantum computing [4–9]. Several excellent reviews of the field are available [10–15].

The fundamental challenges associated with lanthanide ions, concerning their theoretical description and experimental investigation, have been well established for decades [16,17]. After the seminal discovery of slow magnetic relaxation and quantum tunneling of the magnetization in the archetypical SMM Mn_{12}acetate [18,19], research on SMMs and molecular nanomagnets focused mainly on clusters containing transition metal ions. Nevertheless, the potential of incorporating lanthanide ions was soon realized. A striking example, which emerged in this period of research, is the $LnPc_2$ series of single-ion SMMs [20]. However, maybe not surprisingly, researchers largely shied away from the complexities brought in by lanthanide ions for nearly two decades. The situation changed fundamentally when it was realized that with transition-metal based SMMs the blocking temperature is not likely to be further raised substantially [21]. Further work on lanthanide-based molecular clusters followed and indeed showed novel, spectacular properties [1,3,5,6,11]. Focus shifted to the lanthanide systems, and the intense efforts have resulted in remarkable progress and achievements; this special issue is a testimony to it. However, the inherent challenges encountered in lanthanide-containing molecules, of theoretical, experimental and fundamental nature, have essentially not yet been overcome.

In the first part of this work we will discuss these challenges, addressing some aspects, which, in our opinion, deserve larger attention, without attempting to be comprehensive, as excellent complementary reviews are available [22–25]. Our emphasis is on spectroscopic techniques and neutron scattering (NS) in particular. In addition, the considerations are directed towards exchange-coupled poly-nuclear lanthanide-based compounds. We will only briefly comment on single-ion SMMs, since, in our opinion, here the advantages of NS often will not compensate for its disadvantages in comparison to other available experimental techniques.

The NS techniques have seen tremendous progress in the last decade. Throughout the world, long-term programs have been put into place to enhance NS spectrometers and explore novel NS measurement techniques. The development can thus be safely extrapolated to continue at a similar pace for the next decade. Elaborating on the current and future perspectives of NS in our research field may thus be timely, especially as only very few NS studies on lanthanide-based molecular clusters were undertaken to date [26–35].

A frequently cited difficulty with lanthanide ions is their weak exchange coupling, in comparison to what is typically found in transition metal clusters [3,12,14,15]. Indeed, according to the principles for achieving "good SMMs" with high blocking temperature derived from the studies on transition metal-based SMMs, this represents a challenge. However, in our opinion, this aspect is overstressed, since it is not a fundamental limit, and can be overcome by "better" principles. Creating single-ion SMMs is such a principle, and these indeed currently hold the world-record in terms of relaxation barrier [3]. Enhancing the apparent interaction between the lanthanide ions by incorporating non-4f magnetic electrons would be another, exploited in the family of compounds studied in this work. In addition, mixed 3d-4f clusters might deserve more attention, encouraged by the fact that nowadays essentially all hard magnets of technological relevance contain rare earth ions [36]. We will argue that the low symmetry at the lanthanide site usually found in poly-nuclear clusters poses a greater challenge, in terms of the theoretical and experimental characterization. This additional complication may not be favorable for achieving SMMs with high T_B [3,37], but might enable other peculiar magnetic phenomena [4].

In the second part of this work, as a working example, we report original results of a study designed to spectroscopically extract information on the magnetic interactions in the high-T_B $Ln_2(\mu\text{-}N_2{}^{3-})$ system, with Ln = Tb, Dy. The obstruction of weak magnetic coupling between magnetic moments on the 4f electrons has been overcome using a radical $N_2{}^{3-}$ bridge between the lanthanide ions [1,2,38]. In contrast with their non-radical-bridged parent compounds $Ln_2(\mu\text{-}N_2{}^{2-})$, this procedure results in SMMs with the highest blocking temperatures observed so far in a poly-nuclear SMM

(T_B = 14 K in the $Tb_2(\mu-N_2^{3-})$ system) [1]. While the qualitative evidence for the enhanced exchange interactions is present in the low-temperature magnetization data, the quantitative description of this effect is limited to the non-SMM Gd compound (Ln = Gd) based on the isotropic $S = 7/2$ spin of the Gd^{III} ion. The INS technique can offer unique insight into this problem, because excitations based on the exchange interactions are not forbidden by selection rules and can be directly obtained. The INS experiments were conducted on three members of this series, the parent compound $Tb_2(\mu-N_2^{2-})$ (1), the SMM compound $Tb_2(\mu-N_2^{3-})$ (2), and the analogue $Y_2(\mu-N_2^{3-})$ (3), using the spectrometer LET at the ISIS neutron spallation source (Rutherford Appleton Laboratories, Didcot, UK) [39].

The study sheds light on the mentioned aspects. For one, this family of compounds presents an example of how to defeat the weak exchange situation. Secondly, the LET spectrometer represents a latest-generation NS spectrometer and is an example of the dramatic progress in NS mentioned before. Exploiting the time structure of the neutron pulses generated by the ISIS neutron spallation source allowed us, to put it simply, to measure the neutron spectrum for three considerably different incident energy and resolution configurations simultaneously in one run. With traditional spectrometers, one would have to undertake three measurement runs, taking approximately three times longer. This approach obviously has great potential, and the present study represents one of the first efforts to exploit it for a molecular magnetic compound [40,41]. Within this comprehensive work, we have been able to extract a meaningful physical picture for the magnetic ground state of the parent compound $Tb_2(\mu-N_2^{2-})$. A satisfactory description of the SMM compound $Tb_2(\mu-N_2^{3-})$ was not, however, possible because of the intrinsic lack of data in relation to the size of the possible parameter set.

2. General Challenges in Studying the Magnetism in Ln-Based SMMs

2.1. Experimental Aspects of Ln-Based Clusters

To set the stage, let us first comment on mono-nuclear Ln-based clusters and single-ion SMMs in particular. In these systems, the trend is clearly towards molecules with high local symmetry on the lanthanide site, since this has been identified to be crucial for enhancing the SMM property [3,37]. Only in that way "pure" ligand field levels are obtained and for example ground state tunneling can be minimized. Accordingly, the theoretical description of the experimental results by means of phenomenological models is much simplified, as the number of free parameters is much reduced. For instance, the spectroscopic data for $(NBu_4)[HoPc_2]$ and $Na_9[Tb(W_5O_{18})_2]$ could be described with 3 Stevens parameters [30,34]. The proper experimental characterization of such compounds can be a huge challenge, as the example of the $LnPc_2$ molecules shows, but the general approach essentially falls back to an extension of what has been established decades ago.

Given the $\Delta M_J = \pm 1$ selection rule [23,42] in photon-based spectroscopy (electron paramagnetic resonance (EPR), far infrared (FIR), optical, etc.), the high symmetry typically results in few allowed transitions. This is welcome, since it simplifies the analysis, but may also result in silence, for instance in the EPR spectrum. From the perspective of the observability of transitions, low symmetry environments are preferred, since the mixing of states enables more transitions to acquire finite intensity. However, here the spectra often became very complicated, especially in high-resolution techniques such as EPR, which can yield very detailed information that is difficult to extract [25].

For mono-nuclear compounds, INS is governed by the very same selection rule, and thus does not offer any fundamental advantage over the photon-based methods. INS can be, of course, very helpful in obtaining information on ligand-field levels, as it allows one to cover the relevant energy range, and does so in zero magnetic field, which avoids complications. However, there are also significant down-sides, such as low scattering intensity, resolution, absorption and background contributions (vide infra). A further, major obstacle is that INS spectrometers, and NS techniques in general, are not available in-house.

In contrast to the mono-nuclear case, NS techniques do, however, provide additional fundamentally different information when applied to poly-nuclear clusters, which are in the focus in this work. According to the common wisdom typically presented when comparing photon-based and neutron-based spectroscopies, and INS and EPR specifically, INS offers the distinct advantage of a direct observation of exchange splitting, thanks to the INS selection rule $\Delta S = \pm 1$, while these transitions are forbidden in EPR (since here $\Delta S = 0$, where S refers to the spin angular momentum) [22,23]. While these selection rules, of course, apply also to the case of lanthanides, the conclusion as regards the observation of exchange splitting cannot be upheld. A striking recent example is the observation of the exchange splitting in the $[Dy_2(hq)_4(NO_3)_3]$ molecule using EPR techniques [43].

The fundamental advantage of NS over photon-based techniques is its ability to detect spatial distributions and correlations through the dependence of the NS intensity on the momentum transfer, $\mathbf{Q}$. This allows us to extract information from the data, which is not accessible to photon-based spectroscopic methods, since here $\mathbf{Q}$ is practically zero, except when x-ray frequencies are reached. The distinction between NS and (non X-ray) photon-based spectroscopy is thus better cast in terms of the momentum transfer [23], which for NS is typically in the range of $Q = 0.1$–5.0 Å^{-1} (for cold neutron spectrometers), and $Q \approx 0$ for the photon techniques. In view of that, our distinction between mono-nuclear and poly-nuclear systems appears natural.

The greater flexibility given by the INS selection rules implies that more transitions can be observed than in the photon-based methods. In general this is much appreciated, but it also can lead to ambiguities. Although not on a lanthanide-based SMM, the work on NEt$_4$[Mn$^{III}_2$(5-Brsalen)$_2$(MeOH)$_2$Os^{III}(CN)$_6$] provides a text-book example [44]: The INS spectra and magnetization data could be convincingly interpreted within an Ising-exchange model, but was found to be inconsistent with THz-EPR spectra, which were recorded subsequently. Only through the combination of all three techniques, explicitly exploiting the different selection rules for INS and EPR, the three-axis anisotropic nature of the exchange interaction was identified.

Poly-nuclear clusters with low site symmetry should also, in principle, allow richer spectra to be observed than in high-symmetry single-ion molecules. Nevertheless, SMMs based on lanthanide ions can pose a challenge with regard to experimentally obtainable relevant quantities. Essentially, the amount of data that reflect the interaction between the magnetic moments is small, as compared to the number of parameters to be determined in phenomenological models.

Finally, we shall comment on the experimental challenges specific to NS. The complications due to the huge incoherent background produced by the hydrogen atoms in the samples, as well as the relatively low scattering intensity of NS (especially INS), and thus the large required sample masses, are widely recognized [22,23]. The use of lanthanides adds some further complications.

In contrast to the case of 3d metals, some of the lanthanide ions exhibit a large absorption cross section for natural abundance. A comparison for some frequently encountered elements is shown in Table 1. Generally the absorption is somewhat larger than for the transition elements, but Dy, Sm, and especially Gd stand out. NS experiments on Dy compounds are possible but difficult, while they are generally infeasible for Gd compounds. This problem can be bypassed by using low-absorption isotope enriched samples of those elements. For instance, ^{163}Dy and ^{160}Gd have been successfully employed in obtaining spectra [45,46].

Table 1. Neutron absorption cross sections [in units of barns] for some metal elements for natural abundance [47].

H	Cr	Mn	Fe	Y	La	Nd	Sm	Gd	Tb	Dy	Ho	Er	Yb
0.33	13.3	2.6	3.1	1.3	9	50	5922	49,700	23	994	65	159	35

The NS intensity results not only from the magnetic moments in the sample but also from the lattice of nuclei. INS data for instance thus also contain vibrational excitations of the molecule, which need to be distinguished from the magnetic spectrum. This problem seems to be more prevalent in lanthanide containing clusters than in the transition metal clusters. This point can be addressed in several ways, for instance by a Bose correction of high temperature data to estimate the lattice contribution, by performing the same INS experiment on analogue compounds, or substituting for example hydrogen to shift the vibrational frequencies [23,29,32,48].

All the mentioned challenges apply to the $Ln_2(\mu\text{-}N_2{}^{n-})$ compounds investigated in this work. In addition, these compounds are highly air sensitive, which makes them more difficult to handle experimentally, and required special precautions in the planning and undertaking of the experiments.

2.2. Challenges of Analysis

A further difficult intrinsic problem relates to the modelling of poly-nuclear lanthanide-based SMMs. Generally, the modelling is based on effective Hamiltonians containing parameters that need to be determined from experiment, or ab initio calculations (or combinations of both, as for example in the two-step CASSCF approach) [11,49].

A typical effective Hamiltonian for describing the ligand-field levels of a single lanthanide ion is composed of the Stevens operators. The low symmetry of the lanthanide site in principle requires 27 Stevens operators for describing the local anisotropy of the magnetic moment, with the same number of fit parameters (not counting the minor reduction resulting from proper standardization [25]). Notably, already in this step substantial (yet reasonable) assumptions have been made; for describing for example the $J = 15/2$ multiplet for a Dy^{3+} ion, the number of required parameters is actually 119. In addition to the ligand field parameters, terms also need to be added to the effective Hamiltonian to describe the exchange interactions. In a first attempt, when the single-ion J multiplets are considered, these often can be approximated by isotropic Heisenberg exchange [50,51], but for high accuracy also anisotropic/antisymmetric exchange components are required. Therefore, for lanthanide-containing clusters the experiments typically yield less information, while the number of phenomenological parameters is enormously increased, as compared for example to the situation in 3d-only clusters. One obvious way out of this is to consider lower-level effective Hamiltonians, which aim at describing a smaller set of states. This can be successful for describing low-temperature properties, but inevitably fails for understanding the magnetic susceptibility, or the relaxation properties of SMMs [11,37,52]. Alternatively, semi-empirical models such as the point-charge model or improved versions of it [17,24,53] can be used, which promise fewer parameters, but introduce hard to control approximations. They thus typically need to be "calibrated" by a large data set, which may not be available [24].

Ab initio calculations have improved dramatically in recent years and have proven indispensable for arriving at a deep understanding of the electronic structure in the lanthanide-based molecules [11,49]. The calculated results are impressive, yet, usually they do not match the experiments perfectly, leaving room for improvement [30,35]. However, due to the parameter-free nature of these calculations, it is far from clear which tuning knobs would need to be adjusted in order to improve the agreement with experiment. For instance, the ab initio result for the ligand field levels of a specific ion in the cluster in principle can be (and in fact have been) expressed in terms of the Stevens formalism, yielding precise values for all 27 Stevens parameters [32,49]. However, the question arises, which of them should be adjusted and how in order to better match the experimental data.

The situation is this: The effective Hamiltonian approach, which so successfully allows us to bridge the gap between experiment and (ab inito) theory, reaches its limits, as is illustrated in Figure 1. The primary culprit for the issues is the low symmetry at the metal sites, in combination with a lack of a (theoretical) understanding of the relative importance of ligand-field parameters. The latter point prevents experimentalists from choosing minimal yet sensible combinations of parameters in their effective Hamiltonians, and work aimed at overcoming this would, in our opinion, open a path for improving the situation.

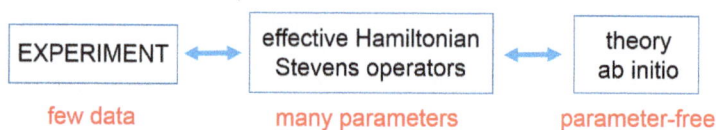

Figure 1. Sketch of the interconnection of challenges in the experimental studies of lanthanide-based systems (for details see text).

2.3. Perspectives of Neutron Scattering Techniques

The lanthanide (Ln^{III}) ion chemistry enables careful studies of entire families of compounds with the same ligand environments. The ligand fields are little affected by chemical substitution and ligand field parameters, when corrected with for example the Stevens parameters, should be largely transferable within a family. This long-known approach has been exploited for instance in inferring the ligand field in the $LnPc_2$ family from NMR and magnetization data [20]. It should be also suitable for systematic NS studies.

We suggest that NS studies on single crystals of molecular magnets should become more commonplace in the future. When using single crystals, INS allows mapping of the full scattering cross section $S(Q,\omega)$, bringing a new light to spin-spin correlations in these materials [54–56]. Similar arguments apply to other NS techniques. In fact, the modern research in quantum magnetism would not be possible if it would not be accompanied by strong efforts in crystal growing. While the necessary tools from the experimental side are present, the main challenge is on the chemists' side: hence, we call for effort to be invested in production of larger single crystals. Such efforts have indeed become accepted as a scientific necessity in the field of quantum magnetism, and we hope they will also become more accepted in our field of research.

The scattering of polarized neutrons is sensitive to both the magnetic nature of the sample, as well as to the directions of its magnetic moments. This experimental fact has been used for a long time to map magnetization densities, for example in magnetic clusters [57,58], and to solve difficult magnetic structures in extended, magnetically ordered systems. Recently, polarized neutron diffraction was applied to probe local anisotropy axes in single-crystal samples of the highly anisotropic transition metal clusters [59,60], leading to a better understanding of the interplay between the ligands and the magnetic properties. This technique is also applicable to lanthanide containing clusters, as well as even more involved polarized NS techniques, such as polarized inelastic neutron scattering.

More parameters are also available in the sample environment. While exchange can be determined using INS without the application of the magnetic field, unlike in many other techniques (e.g., EPR), magnetic fields of up to 17 T are standardly available on neutron sources. Neutron scattering samples can also be placed into pressure cells, and submitted to uniaxial or hydrostatic pressures [23].

All the mentioned techniques and approaches are going to benefit significantly from the availability of new generations of sample environments, such as for example the recently constructed 26 T magnet in the Helmholtz Zentrum Berlin, more advanced instruments, for example LET, as well as the suite of instruments planned to be constructed at the high-flux European Spallation Source (ESS). This will allow for smaller samples, more extreme conditions, systematic studies of larger sample families, and will lead to higher throughput of experimental results. The new developments are going to benefit the neutron scattering community as well as the molecular magnetism field as a whole.

3. Inelastic Neutron Scattering Study of the $Tb_2(\mu$-$N_2)$ System

3.1. Introduction to the $Tb_2(\mu$-$N_2)$ System

The compound $[K(18$-crown-$6)(THF)_2][\{[(Me_3Si)_2N]_2(THF)Tb\}_2(\mu$-$\eta^2{:}\eta^2$-$N_2)]$ (**2**), or $Tb_2(\mu$-$N_2^{3-})$ in shorthand, shows SMM behavior with a blocking temperature of ~14 K [1]. It is derived from a parent compound $\{[(Me_3Si)_2N]_2(THF)Tb\}_2(\mu$-$\eta^2{:}\eta^2$-$N_2)$ (**1**) [61], or $Tb_2(\mu$-$N_2^{2-})$ but differs by having one fewer electron on the dinitrogen bridge. In addition, $[K(18crown-6)(THF)_2]^+$ cations are present in the crystal lattice of **2**, which will be of importance in what follows. The family of compounds also includes the Dy^{III}-containing molecules $Dy_2(\mu-N_2{}^{2-})$ and $Dy_2(\mu-N_2{}^{3-})$, the Gd^{III}-containing molecules $Gd_2(\mu-N_2{}^{2-})$ (**4**) and $Gd_2(\mu-N_2{}^{3-})$ (**5**), and the Y^{III} analogue $Y_2(\mu-N_2{}^{3-})$ (**3**) [2].

Figure 2 shows the molecular structures of the parent and derived SMM molecules **1** and **2**. The cores of **1** and **2** consists of two Tb^{III} ions ($J = 6$, $g_J = 1.5$) coupled via dinitrogen bridges $N_2{}^{2-}$ and $N_2{}^{3-}$, respectively. In both compounds, the Tb sites are occupying a crystallographically equivalent but low-symmetry site. The additional electron on the dinitrogen bridging unit in the SMM compound **2** is considered to increase the magnetic coupling strength significantly [1,2]. Indeed, fits to the magnetic susceptibility of the Gd^{III} compounds **4** and **5** yielded coupling strengths of $\mathcal{J} = -1.4$ K and $\mathcal{J} = -78$ K (in $\mathcal{J}$ notation), respectively, as well as evidence for a weak intermolecular interaction of $\mathcal{J}'$ in **5** [1]. These compounds are not suitable for INS studies due to the large neutron absorption cross sections for natural Gd, as discussed above.

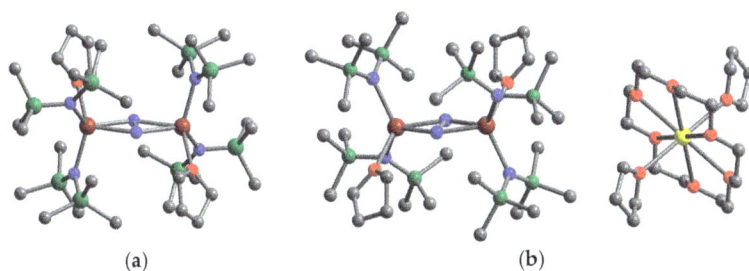

(a)   (b)

Figure 2. (a) Molecular structure of the parent compound $Tb_2(\mu-N_2{}^{2-})$ (**1**); (b) Molecular structure of the SMM (single-molecule magnet) compound $Tb_2(\mu-N_2{}^{3-})$ (**2**). In both panels: Tb^{III} in dark red, N in blue, O in light red, Si in green, C in gray, K in yellow, H atoms were omitted.

The molar magnetic susceptibilities of the parent and SMM compounds **1** and **2** were reported previously [1]. The magnetic susceptibility of the parent compound **1** is shown in Figure 3a. The χT vs. T curve grows monotonically from a low value of 3.4 cm^3K/mol at the lowest temperature of 2 K and flattens out at high temperatures approaching the Curie value of 23.62 cm^3K/mol. An overall down turn of the χT curve with lowering temperature is typical for ligand-field levels of lanthanide ions, but for Tb^{III}, the curve should approach a significant finite value at zero temperature in a pure ligand field model [16,17]. The drop to nearly zero at the lowest temperatures is consistent with a weak antiferromagnetic exchange interaction between the Tb^{III} magnetic moments.

The molar magnetic susceptibility χT vs. T of the SMM compound **2**, for temperatures above its blocking temperature, is shown in Figure 3b. At 300 K the χT value is 22.9 cm^3K/mol. As the temperature is lowered, the susceptibility grows, which is expected for the effective ferromagnetic alignment between the Tb^{III} magnetic moments. The data show a broad maximum at about 70 K, reaching a χT value of 34.6 cm^3K/mol, followed by a decrease at lower temperatures, with $\chi T = 31.0$ cm^3K/mol at 15.6 K. The down turn could suggest the presence of excited states in the energy range of ca. 70 K with higher magnetic moment than the ground state, which get depopulated at low temperatures. An alternative could be the presence of weak antiferromagnetic intermolecular interactions (vide infra).

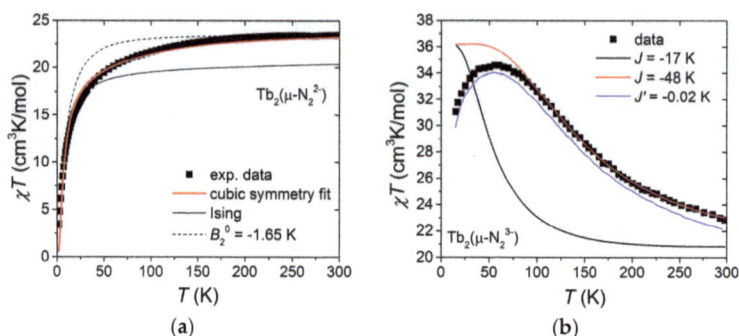

Figure 3. (a) Molar magnetic susceptibility data (squares) of the parent compound **1** collected at 1 T and the calculations (lines) based on the three models discussed in the text; (b) Molar magnetic susceptibility of the SMM compound **2** (squares) and the calculations (lines) based on several models discussed in the text.

3.2. Experimental Details

In order to determine the thermodynamic magnetic behavior in the ground state of the parent compound **1**, field-dependent magnetization curves were recorded. The maximum field was 7 T, and temperature ranged from 2 K to 20 K.

In view of the expected challenges with studying and analyzing the magnetism in the SMM compound **2**, as described previously, it is fortunate that the parent compound **1** and the analogue with diamagnetic Y^{III}, **3** are also available, as each can yield important insights into the vibrational background and the exchange couplings in the SMM complex in **2**. The INS experiments, using the LET spectrometer at the ISIS facility, were therefore conducted on all three compounds. Regarding the comparison of results, it should be noted, however, that the vibrational spectrum for the parent compound **1** can be expected to be very different from those for the compounds **2** and **3**, due to the presence of the K-crown cations in the latter. In addition, the additional charge on the dinitrogen bridge in **2** should significantly affect the ligand field at the Tb^{III} sites in this compound. The ligand fields in **1** and **2** are thus not comparable, which must not be overlooked.

The INS spectra were measured in three energy ranges with incident neutron energy of 2 meV (low-energy range), 11 meV (intermediate-energy range), and 22 meV (high-energy range). Positive energies refer to neutron energy loss. The temperatures were varied from the base temperature of 2 K to 100 K, in several steps. The data permitted analyzing the full $S(Q,\omega)$ plot. The integrated INS intensity as a function of energy is shown for selected measurement conditions; some additional results are presented in the SI.

3.3. Magnetization Data for the Parent Compound **1**

The low-temperature magnetization data for the parent compound **1** are shown in Figure 4. At 2 K, the magnetization displays an inflection point at about 1 T and then grows rapidly until about 5 T, but does not fully saturate even at the maximum field of 7 T. The higher temperature data gradually wash out the low-field inflection feature and display an even bigger obstacle to saturation. The low-field inflection point is an indication of weak antiferromagnetic exchange interactions between the two Tb^{III} ions in the cluster. For an isolated $\pm M_J$ doublet, the powder-averaged saturation magnetization is calculated to be approximately $1/2\mu_B g_J M_J$ or ~ 9 μ_B for Tb^{III} and $M_J = J = 6$. The observed maximum magnetization at 7 T of 9.21 μ_B thus strongly suggests a $M_J = \pm 6$ doublet for the Tb^{III} ground state. This finding is consistent with the expectation from electrostatic considerations [10].

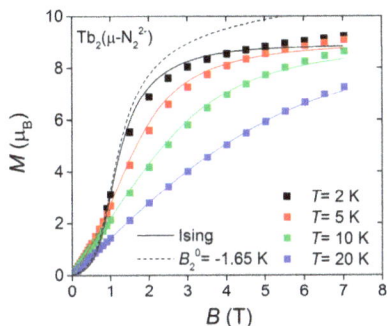

Figure 4. Magnetization data (squares) at different temperatures for the parent compound **1** and the calculations (lines) based on two models discussed in the text. The colored solid lines represent the results for the Ising model at temperatures of 2, 5, 10, 20 K (black to blue). The dashed line represents the result for Equation (2) at 2 K.

3.4. Inelastic Neutron Scattering Data for the Parent Compound **1**

Figure 5a shows the temperature dependence of the low-energy INS spectrum collected for the parent compound **1**. The main feature is a clear excitation at 0.75 meV (peak I). Its intensity decreases at higher temperatures on the neutron energy-loss side, and shows the corresponding temperature dependence on the neutron energy-gain side, which is typical for a cold magnetic transition. In addition, this peak is present at low momentum transfer Q, which rules out a phononic origin (see Figure S1). Thus, peak I, and its anti-Stokes companion peak I′, can be unambiguously assigned to a cold magnetic transition at 0.75 meV.

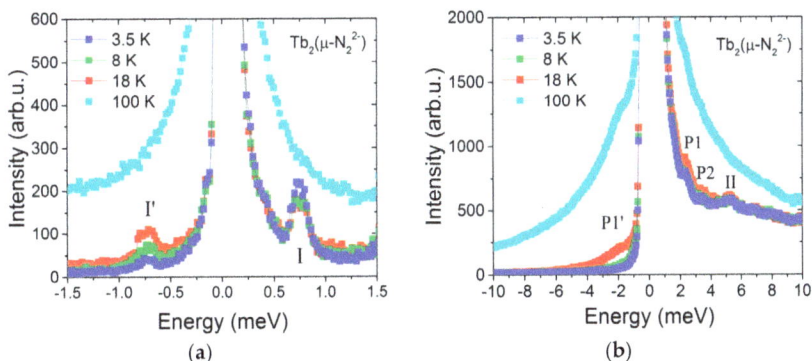

Figure 5. (a) Low-energy INS spectrum in the parent compound **1**. Peak I indicates the exchange-based transition and I′ its anti-Stokes pair; (b) Intermediate-energy levels in the parent compound **1**. Peaks P1 and P2 denote vibrational levels with P1′ the anti-Stokes pair of P1. Peak II indicates a ligand-field transition at 5.2 meV.

The intermediate-energy data shown in Figure 5b display additional levels at about 2 meV (peak P1), 3 meV (peak P2) and 5 meV (peak II). Based on the temperature dependence, only peak II behaves as a magnetic transition, which could be cold or emerge from a possibly very-low lying excitation. The intensity of this transition is large even at low Q, which is further strong evidence for a magnetic origin of the peak (see Figure S2). Based on the temperature and Q dependence, the 2 meV and 3 meV transitions are assigned to lattice vibrations (since, for example, the 2 meV transition grows on both sides with temperature).

No additional magnetic peaks could be identified in the high-energy data. From the INS data, the presence of two cold magnetic transitions in **1** is thus concluded, at 0.75(2) meV (peak I) and 5.2(2) meV (peak II).

3.5. Inelastic Neutron Scattering Data for the SMM Compound **2** and Y^{III} Analogue **3**

Figure 6a shows the intermediate-energy range INS data at base temperature for the SMM compound **2**, together with the data for its analogue containing diamagnetic Y^{III} centers, **3**. There are several peaks in this energy range. However, comparing the data of **2** with that of compound **3** enables the exclusion of most of the observed spectrum as vibrational. In **2**, there is one clear excitation at ~9 meV (peak I), which is not present in compound **3**, and can hence be assigned to a magnetic origin. There is an additional candidate for a magnetic transition at ~8.5 meV (indicated by the question mark), but if it exists it coincides with large vibrational background peaks. With the present data it cannot be identified unambiguously.

Figure 6b presents the measured temperature dependence for compound **2**. The intensity of peak I decreases at higher temperatures, which is a clear signature of a cold magnetic transition. This peak could not be seen well in the $S(Q,\omega)$ plot due to its low intensity, and thus no conclusions concerning its origin could be drawn from its Q dependence. Additionally, Bose corrections did not yield good estimates of the backgrounds (see Figure S3).

Further magnetic scattering intensity could not be identified in either the low-energy or the high-energy data. The INS experiments performed on **2** thus provide evidence for one cold magnetic transition at 9.2(2) meV (peak I). The existence of this transition plays a discerning role in the analysis below. However, the experimental evidence is, admittedly, not extremely strong. For that reason the available INS data were analyzed repeatedly with the greatest care, and it was concluded that it is of magnetic origin, but a word of caution is appropriate.

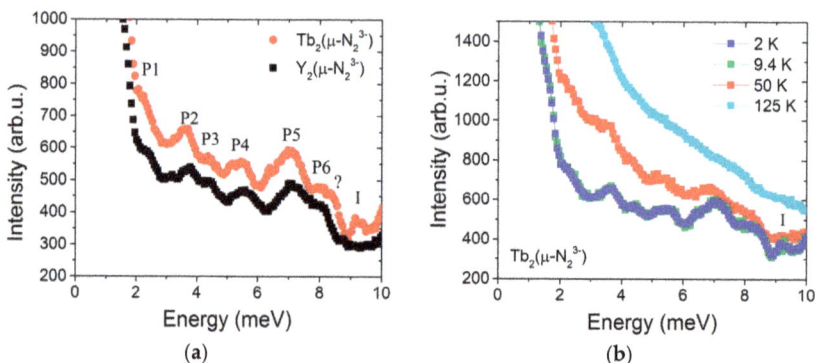

Figure 6. (a) Intermediate-energy INS data at 2 K in the SMM compound **2** (red) compared to the vibration spectrum in the analogue containing diamagnetic Y^{III}, **3** (black). The peaks labelled P1–P6 denote vibrational excitations seen in both compounds. The peak I at 9.2 meV is indicated; (b) Intermediate-energy INS spectra measured for **2** at different temperatures. Note the offset on the y axis in these plots, demonstrating a large incoherent scattering background.

4. Discussion

4.1. Insights from the Point Charge Model

In order to gain understanding of the single-ion properties of the investigated systems, a set of point-charge model calculations [17,53] were performed. Importantly, this simple model was not used as a quantitative device for accessing exact parameters of the local Hamiltonian. In contrast, we sought to obtain generic information about the spectra and the single-ion wave functions for qualitative results as the low symmetry of the Tb^{III} site makes the problem intractable. For this purpose, the Tb^{III} environment was first approximated by a tetrahedral charge environment as shown in Figure 7, with two of the charges variable (representing the N_2^{n-} bridge, with $n = 2$ or 3, and the difference between the oxygen and nitrogen ion charge).

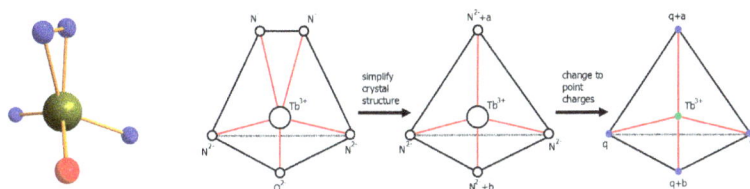

Figure 7. The local low-symmetry environment of the Tb^{III} ion and its reduction to an approximate two-parameter point-charge model, which captures the most relevant generic aspects.

The generic result of this procedure is shown in Figure 8. The Tb^{III} ion, surrounded by a polar environment, displays a non-Kramers doublet spectrum, with an approximate $M_J \approx \pm 6$ ground state, followed by an excited $M_J \approx \pm 5$ doublet ("doublet" is henceforth used to denote a non-Kramers doublet). The dominant components of the single-ion wave function pair are in the $M_J = +6$ and $M_J = -6$ sectors. However, there are small contributions to the other M_J components, which are essentially given by the symmetry of the ion's environment. For example, in a polar tetrahedral environment ($b = 0$, $a < 0$ in Figure 7), the ground state contains small $M_J = \pm 3$ and $M_J = 0$ components, as shown in Figure 8b. In the case of a low symmetry for the Tb^{III} ion, as in the studied compounds, *all* of the single-ion components have finite values, albeit much smaller than the dominant component.

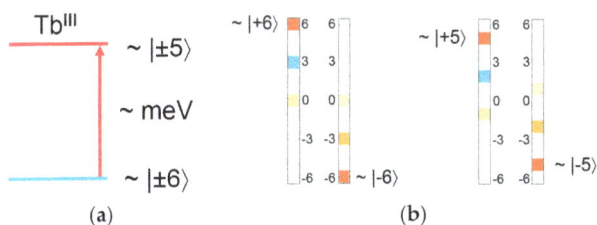

Figure 8. (a) The typical single-ion low energy spectrum with non-Kramers doublet single-ion wave functions of the Tb^{III} ion coming from the approximate ligand environment discussed in the text; (b) The bars to the right and left represent the wave functions of the $M_J \approx \pm 6$ and $M_J \approx \pm 5$ doublets in a polar tetrahedral environment, respectively, with the magnitude of the individual M_J components colour coded (red = 1, white = 0, blue = −1). In lower symmetry, the "white" components would all gain finite values.

This is an important observation for neutron scattering: The $\Delta M_J = \pm 1$, 0 INS selection rule permits INS transitions between the $M_J \approx \pm 6$ ground and $M_J \approx \pm 5$ excited ligand field states, but it would result in zero INS scattering intensity for exchange-split states if the levels were pure M_J states, as in Ising exchange models. As will be shown in detail below, if the exchange is of Ising-type, then the excitations resulting from the exchange interaction correspond to spin flips with a large associated change of the z component of the magnetic moment J^z, or M_J in fact. For instance, a transition involving a spin flip from $M_J = -6$ to $M_J = +6$ emerges, for which $\Delta M_J = 12$. However, since there are non-zero components of the initial and final states that produce $\Delta M_J = \pm 1$, 0 overlaps, it is possible to observe weak intensity in INS corresponding to these exchange-split transitions.

A further generic result of the point-charge investigation is that the lowest excitation is several meV above the ground state, and that the additional charge on the radical bridge in the SMM compound strongly shifts the ligand-field levels to even higher energies. For instance, the lowest excitation shifts from a ~6 meV range to a ~60 meV range. In other words, the magnetic system is expected to become much more anisotropic and Ising type as the competing states are pushed further away in energy. Hence, we expect that for the description of low-temperature thermodynamic quantities, we may restrict ourselves to the ground state doublet of the system, especially in the SMM compound **2**.

4.2. The Parent Compound **1**

The parent compound is described in terms of a Heisenberg spin Hamiltonian:

$$H = -\mathcal{J}\, J_1 \cdot J_2 + \sum_{i=1,2} \sum_{k=2,4,6} \sum_{q=-k}^{k} B_k^q O_k^q(i) \tag{1}$$

Here, the first part describes the usual exchange interaction between the two Tb^{III} ions, and the second part describes all the possible contributions to the ligand field in terms of the Stevens operator formalism [16,17,53]. The exchange interaction between the J multiplets of lanthanide ions can generally be well described by isotropic Heisenberg exchange [50]. Due to the large magnetic moments and weak exchange in lanthanide ions, dipolar interactions can also be appreciable [16]. These are neglected here also, because their effects are similar to those of the ligand field terms and difficult to discern. Due to the aforementioned fundamental problems with the quantity of data and the results of the point charge modelling, a much reduced Hamiltonian was also considered:

$$H' = -\mathcal{J}\, J_1 \cdot J_2 + \sum_{i=1,2} B_2^0 O_2^0(i) \tag{2}$$

The uniaxial anisotropy operator $O_2^0(i)$ allows us to mimic the effect of the ligand-field environment on the low-temperature properties of the system. The advantage of this reduction is, of course, that the Hamiltonian H' contains only two parameters.

In case of a strong Ising-type anisotropy or large negative value of B_2^0, the Hamiltonian of the system essentially reduces to a low-temperature dimer model with pure Ising exchange interactions. In Section 3.4 above, the low-temperature susceptibility and magnetization was found to indicate small antiferromagnetic interactions present in the system. The ground state and lowest exchange-split excitations in such an Ising dimer stem from the single-ion $M_J \approx \pm 6$ doublets, as indicated in Figure 9b. The lowest excitation from the ground state corresponds to a spin flip on one site and has an excitation energy of $\Delta E = 72|\mathcal{J}|$. Let us compare the results of this model to the experimentally observed excitations shown in Figure 9a: Association of the observed 0.75 meV magnetic peak with this transition results in $\mathcal{J} = -0.12$ K. Note that also in the Gd^{III} compound **4**, antiferromagnetic intra-molecular interactions were inferred [2], of strength $\mathcal{J} = -1.41$ K, which is qualitatively consistent with our finding for **1**.

In Figures 3a and 4, simulations of the magnetization and susceptibility curves are shown using the determined value of the interaction and Equation (2) with variation of B_2^0. For infinitely large B_2^0, the model reduces to that of a dimer of two-level states with pure Ising interactions and contains only one parameter, namely $\mathcal{J}$. With $\mathcal{J}$ taken from our INS results, this establishes a parameter-free model for the low-temperature magnetism in **1**. The resulting simulations are shown as solid lines in Figures 3a and 4. Remarkably, the measured magnetization is very well reproduced, demonstrating the validity of this model for the ground-state properties of **1**. The susceptibility is well reproduced at low temperatures, but strongly deviates above ~30 K (see Figure 3a). This is expected, since the ligand-field levels that govern the magnetism at higher temperatures are not present in the model (they are shifted to infinite energy by the infinite B_2^0).

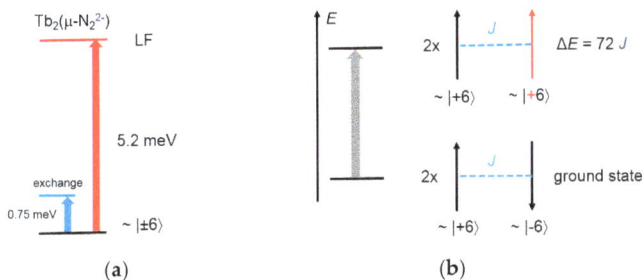

Figure 9. (a) Excitation energy scheme experimentally observed in the parent compound **1**; (b) Theoretically expected excitation spectrum of an Ising dimer formed by two exchange-coupled $M_J \approx \pm 6$ doublets. A weak INS transition occurs due to the small M_J components in the involved states, as discussed in the text.

In the next step, the value of B_2^0 was thus chosen such that Equation (2) reproduces the observed excitation at 5.2 meV, yielding $B_2^0 = -1.65$ K. The simulated susceptibility curve (dashed line in Figure 3a) now correctly approaches the Curie value at high temperatures, but otherwise reproduces the data poorly, showing χT values that are too large in the intermediate temperature range of ~70 K. In addition, the description of the high-field part of the magnetization is worse (dashed line in Figure 4). Obviously, the magnetic contribution of the first excited ligand-field level at 5.2 meV is significantly overestimated in this model.

It is possible to obtain a relatively good fit to the magnetic susceptibility data using an extended set of Stevens operators in addition to the exchange. The red curve in Figure 3a was calculated assuming an approximate local cubic environment, with $B_2^0 = -640(27) \times 10^{-2}$ K, $B_4^0 = -77(11) \times 10^{-4}$ K, and $B_4^3 = -84(32) \times 10^{-4}$ K. However, this by no means was the only reasonable fit we found. In fact, similar fits were obtained with substantially different sets of Stevens parameters, which underpins the well-known challenges with over-parametrization in the fitting of experimental susceptibility curves. The lowest ligand field levels expected from these fits occur at around 25 meV, much larger than observed 5.2 meV peak, pointing again to the low magnetic moment associated with this excitation.

4.3. The SMM Compound **2**

The main difference, from the view point of magnetic modeling, between the parent compound **1** and the SMM compound **2** is that the magnetic exchange acts via a $s = 1/2$ electron spin on the radical dinitrogen bridge, which changes the form of the Hamiltonian to:

$$H_{SMM} = -\mathcal{J} \left(J_1 \cdot s + s \cdot J_2 \right) + \sum_{i=1,2} \sum_{k=2,4,6} \sum_{q=-k}^{k} B_k^q O_k^q(i) \tag{3}$$

An exchange directly between the Tb^{III} ions is not included, since it can be safely assumed to be much smaller than the exchange to the radical spin and showed negligible effects in test simulations. Again, based on similar arguments as before, a simplified model of the system is considered:

$$H'_{SMM} = -\mathcal{J} \left(J_1 \cdot s + s \cdot J_2 \right) + \sum_{i=1,2} B_2^0 O_2^0(i) \tag{4}$$

In a situation with large Ising-type anisotropy (B_2^0 very large), one expects that, in the ground state, the Ising-like moments of the Tb^{III} ions remain parallel. If the interaction $\mathcal{J}$ is antiferromagnetic, then in the ground state the radical spin s is essentially antiparallel to the Tb^{III} moments and parallel in the ferromagnetic case. The first excitation of the system corresponds to a spin-flip of a Tb^{III} moment and occurs at an energy of $\Delta E = 6|\mathcal{J}|$. A second excitation emerges at an energy of $\Delta E = 12|\mathcal{J}|$, which is related to a spin flip of the central radical spin. The exchange-split level diagram is depicted in Figure 10b. In our INS data, we observed a single magnetic peak at 9.2 meV. If this were associated with the lowest exchange-based excitation, an exchange constant of $\mathcal{J} = -17$ K would result.

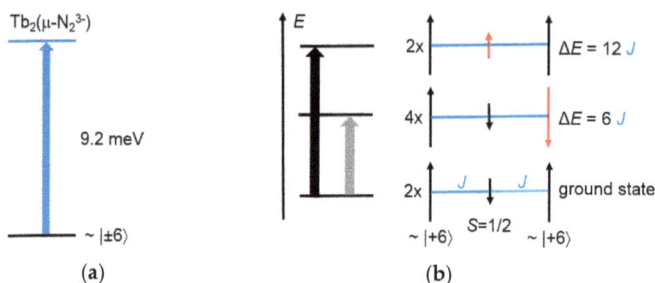

Figure 10. (a) Excitation energy scheme experimentally observed in the SMM compound **2**; (b) Theoretically expected excitation spectrum for the Ising-exchange model of the SMM compound **2** discussed in the text as a basic model. The INS transitions from the ground state to the second excited state is allowed (black arrow). A further weak INS transition from the ground state to the first excited state occurs due to the small M_J components in the involved states, as discussed in the text.

Figure 3b compares the calculation based on this value (black solid line) to the experimental susceptibility data [1]. The agreement is poor due to a significant underestimation of the exchange constant. Indeed, if one assumes an antiferromagnetic exchange interaction about three times larger of $\mathcal{J} = -48$ K (red solid line) one gets fairly good agreement with the experimental curve at higher temperatures. At present, the cause of this discrepancy is unclear. Note that associating the observed 9.2 meV transition with the expected excitation at $\Delta E = 12|\mathcal{J}|$, which could have stronger INS intensity, worsens the situation by another factor of two.

The assumed model is certainly simplified, but based on the generic findings from the point-charge considerations and the overall SMM character of the system at low temperatures, one would expect the Ising-type anisotropic model to hold better than in the parent compound **1**. An exchange-based excitation lower than $6|\mathcal{J}|$ cannot arise in such models. Interestingly, the tripled exchange coupling, $\mathcal{J} = -48$ K, is more consistent with the exchange interaction observed in the related Gd^{III} compound **5**, and moreover predicts a first excitation at 24.8 meV or 290 K, in close agreement with the energy barrier of 330 K inferred from ac susceptibility measurements performed on **2** [1]. On the other hand, this ~25 meV excitation would then indeed be the lowest excitation, and the observed INS peak at 9.2 meV would remain unaccounted for, as well as the observed significant down turn of the magnetic susceptibility at temperatures below ~70 K. The latter would indicate a ground state of the Tb^{III} ions which has a lower magnetic moment than the $M_J \approx \pm 6$ doublet emerging in any model based on a strongly Ising-type anisotropy.

These discrepancies and the decrease of the susceptibility at lower temperatures suggest the possibility of antiferromagnetic intermolecular interactions [2,62]. In a molecular field approach, this scenario yields the susceptibility $\chi = \chi_{SMM}/(1 + \lambda \chi_{SMM})$, where χ_{SMM} is the calculated susceptibility of an isolated Tb$_2$(μ-N$_2{}^{3-}$) unit, which fits the experimental curve remarkably well with $\lambda = 0.06$ mol/emu. In an attempt to establish a more realistic model, we connect the trimeric units into a ladder configuration assuming the intermolecular couplings of $\mathcal{J}'$ only between the Tb^{III} moments. Quantum Monte Carlo simulations using the ALPS framework [63,64] were performed, with the ladder length set to 20 molecules. As shown in Figure 3b, the addition of a small intermolecular interaction of $\mathcal{J}' = -0.02$ K (blue solid line) is able to reproduce the observed low-temperature decline in the magnetic susceptibility. The origin of this effect may be the dipole-dipole interactions between the Tb^{III} moments of neighboring molecules, which are estimated to ~0.05 K, and therefore could account for the required magnitude of $\mathcal{J}'$ [2]. The intermolecular interactions give rise to an associated, nearly dispersion-less excitation at $288|\mathcal{J}'| = 0.5$ meV, which is too low to account for the 9.2 meV excitation seen in INS, and no INS feature was observed at this energy.

5. Materials and Methods

Neutron scattering: Non-deuterated powder samples were synthesized following published procedures [1]. Sample shipment and handling was undertaken very carefully because of the known air sensitivity of the compounds. Cooled and sealed samples were directly shipped to the ISIS facility at the Rutherford Appleton Laboratory in Chilton, UK, and were stored in a freezer at $-40\,^{\circ}$C. Samples were wrapped into aluminum foil and mounted in the standard cans used at ISIS, all within a glovebox. Each sample was prepared shortly before the experiment, and quickly inserted into the Orange cryostat and cooled down. The sample quantities were weighed and found to be 0.834 g for **1**, 0.914 g for **2** and 1.516 g of **3**. Data were collected at the LET time-of flight neutron spectrometer at the ISIS neutron source in a multi-rep mode, in which multiple Q-energy windows can be obtained from a single measurement. Several settings were used in order to obtain an overview in Q-E space. Instrument settings were different by incident neutron energies, chopper speeds and the resulting choice of energy windows. The energy windows with maximum energy transfer of $E = 2.01$ meV, 11.7 meV, 12.5 meV, 17.4 meV and 22.1 meV were used. The most relevant data were obtained with the $E = 2.01$ meV (resolution at the elastic line of 160 μeV) and 11.7 meV (resolution at the elastic line of 500 μeV) windows, which captured the low and intermediate energy excitations in the system. All data were corrected for empty can and vanadium measurements. The data were also scaled by the measured sample weights. The 2.01 meV and 11.7 meV energy scans shown here were obtained by summing up to $Q = 0.7$ Å^{-1} and $Q = 1.8$ Å^{-1}, respectively.

Magnetic Measurements: All handling of sample **1** in preparation of magnetic measurements was executed with a Teflon-coated spatula. A crushed crystalline sample of **1** was loaded into a 7 mm diameter quartz tube and was coated with sufficient eicosane to restrain the sample. The quartz tube was fitted with a sealable hose-adapter, evacuated on a Schlenk line, and then flame-sealed under vacuum. Magnetic susceptibility measurements were performed using a MPMSXL SQUID magnetometer (Quantum Design, Inc., San Diego, CA, USA). Dc magnetic susceptibility measurements were performed at temperatures ranging from 2 to 300 K (variable temperature) at 1 T and the magnetization was measured in fields ranging from 0 to 7 T at fixed temperatures. All data were corrected for diamagnetic contributions from the eicosane and for diamagnetism estimated using Pascal's constants [65].

6. Conclusions

In conclusion, we have discussed the challenges and perspectives of neutron scattering in lanthanide-based molecular magnets. The focus of the discussion was on the poly-nuclear clusters with low local symmetry, which present inherent challenges for an experimentalist: a situation of having little data, but many parameters in the effective Hamiltonian, and stiff, parameter-free ab initio calculations.

In the second part of the paper, we presented original results of an inelastic neutron scattering (INS) study on a high blocking temperature single molecule magnet (SMM), its Y^{III} analogue and a non-SMM parent compound. In the parent compound, we observed two peaks, I at 0.75(2) meV and II at 5.2(2) meV. Together with a simple, but plausible Ising spin model suggested by point-charge calculations, peak I allowed us to clarify the low-energy behavior of the material, notably the single-ion ground states, an approximate $M_J \approx \pm 6$ pseudo doublet, and the exchange interaction $\mathcal{J} = -0.12$ K. The physical picture thus obtained, fits well to the low temperature magnetization data without the need for introducing any further parameters. However, additional intermediate- and higher-energy data are needed to fully describe the system. To the best of our knowledge, this is the first time the exchange interaction between lanthanide ions was *directly* determined based upon INS.

In the SMM compound, we observed, among several phonon peaks, a very weak magnetic excitation at 9.2(2) meV. The assignment of the peak to the exchange-split level within the Ising model results in the exchange interaction value of $\mathcal{J} = -17$ K, which does not reproduce the susceptibility curve. We showed that within the Ising model a larger interaction of $\mathcal{J} = -48$ K is required for this, together with an intermolecular exchange of $\mathcal{J}' = -0.02$ K. The reason for the discrepancy between the INS and susceptibility data is at present not clear, but it points to the necessity for a more complex model and additional data points required for its validation.

Supplementary Materials: The following are available online at www.mdpi.com/2312-7481/2/4/45/s1, Figure S1: Low-energy $S(Q,\omega)$ spectrum of the parent compound **1**; Figure S2: Intermediate-energy $S(Q,\omega)$ spectrum of the parent compound **1**; Figure S3: INS spectra of compounds **1**, **2** and **3** with the lattice contributions estimated by the Bose correction procedure.

Acknowledgments: K.P. and O.W. thank J. Mutschler for help with point-charge model calculations. W.J.E. thanks the U.S. National Science Foundation for support (CHE-1565776). The work performed at University of California, Berkeley was supported by the National Science Foundation (NSF) under Grant CHE-1464841.

Author Contributions: O.W. and J.N. conceived and designed the INS experiments; S.D. and J.R.L. conceived and designed the magnetic susceptibility measurements; J.F.C. and W.J.E. contributed samples; O.W., J.N., T.G. and S.D. performed the experiments; K.P., J.N. and O.W. analyzed the data; K.P. and O.W. wrote the core of the paper and all authors contributed to revising it.

Conflicts of Interest: The authors declare no conflict of interest.

References and Notes

1. Rinehart, J.D.; Fang, M.; Evans, W.J.; Long, J.R. A N_2^{3-} Radical-Bridged Terbium Complex Exhibiting Magnetic Hysteresis at 14 K. *J. Am. Chem. Soc.* **2011**, *133*, 14236–14239. [CrossRef] [PubMed]

2. Rinehart, J.D.; Fang, M.; Evans, W.J.; Long, J.R. Strong exchange and magnetic blocking in N_2^{3-}-radical-bridged lanthanide complexes. *Nat. Chem.* **2011**, *3*, 538–542. [CrossRef] [PubMed]

3. Chen, Y.-C.; Liu, J.-L.; Ungur, L.; Liu, J.; Li, Q.-W.; Wang, L.-F.; Ni, Z.-P.; Chibotaru, L.F.; Chen, X.-M.; Tong, M.-L. Symmetry-Supported Magnetic Blocking at 20 K in Pentagonal Bipyramidal Dy(III) Single-Ion Magnets. *J. Am. Chem. Soc.* **2016**, *138*, 2829–2837. [CrossRef] [PubMed]

4. Ungur, L.; Lin, S.-Y.; Tang, J.; Chibotaru, L.F. Single-molecule toroics in Ising-type lanthanide molecular clusters. *Chem. Soc. Rev.* **2014**, *43*, 6894–6905. [CrossRef] [PubMed]

5. Sharples, J.W.; Collison, D.; McInnes, E.J.L.; Schnack, J.; Palacios, E.; Evangelisti, M. Quantum signatures of a molecular nanomagnet in direct magnetocaloric measurements. *Nat. Commun.* **2014**, *5*, 5321. [CrossRef] [PubMed]

6. Shiddiq, M.; Komijani, D.; Duan, Y.; Gaita-Ariño, A.; Coronado, E.; Hill, S. Enhancing coherence in molecular spin qubits via atomic clock transitions. *Nature* **2016**, *531*, 348–351. [CrossRef] [PubMed]

7. Tang, J.; Hewitt, I.; Madhu, N.T.; Chastanet, G.; Wernsdorfer, W.; Anson, C.E.; Benelli, C.; Sessoli, R.; Powell, A.K. Dysprosium Triangles Showing Single-Molecule Magnet Behavior of Thermally Excited Spin States. *Angew. Chem. Int. Ed.* **2006**, *45*, 1729–1733. [CrossRef] [PubMed]

8. Chibotaru, L.F.; Ungur, L.; Soncini, A. The Origin of Nonmagnetic Kramers Doublets in the Ground State of Dysprosium Triangles: Evidence for a Toroidal Magnetic Moment. *Angew. Chem. Int. Ed.* **2008**, *47*, 4126–4129. [CrossRef] [PubMed]

9. Luis, F.; Repollés, A.; Martínez-Pérez, M.J.; Aguilà, D.; Roubeau, O.; Zueco, D.; Alonso, P.J.; Evangelisti, M.; Camón, A.; Sesé, J.; et al. Molecular Prototypes for Spin-Based CNOT and SWAP Quantum Gates. *Phys. Rev. Lett.* **2011**, *107*, 117203. [CrossRef] [PubMed]

10. Rinehart, J.D.; Long, J.R. Exploiting single-ion anisotropy in the design of f-element single-molecule magnets. *Chem. Sci.* **2011**, *2*, 2078–2085. [CrossRef]

11. Ungur, L.; Chibotaru, L.F. Strategies toward High-Temperature Lanthanide-Based Single-Molecule Magnets. *Inorg. Chem.* **2016**, *55*, 10043–10056. [CrossRef] [PubMed]

12. Woodruff, D.N.; Winpenny, R.E.P.; Layfield, R.A. Lanthanide Single-Molecule Magnets. *Chem. Rev.* **2013**, *113*, 5110–5148. [CrossRef] [PubMed]

13. Tang, J.; Zhang, P. *Lanthanide Single Molecule Magnets*, 1st ed.; Springer: Berlin/Heidelberg, Germany, 2015.

14. Liddle, S.T.; van Slageren, J. Improving *f*-element single molecule magnets. *Chem. Soc. Rev.* **2015**, *44*, 6655–6669. [CrossRef] [PubMed]

15. Sessoli, R.; Powell, A.K. Strategies towards single molecule magnets based on lanthanide ions. *Coord. Chem. Rev.* **2009**, *253*, 2328–2341. [CrossRef]

16. Jensen, J.; Mackintosh, A.R. *Rare Earth Magnetism*; Clarendon Press: Oxford, UK, 1991.

17. Newman, D.J. Theory of lanthanide crystal fields. *Adv. Phys.* **1971**, *20*, 197–256. [CrossRef]

18. Christou, G.; Gatteschi, D.; Hendrickson, D.N.; Sessoli, R. Single-Molecule Magnets. *MRS Bull.* **2000**, *25*, 66–71. [CrossRef]

19. Gatteschi, D.; Sessoli, R. Quantum Tunneling of Magnetization and Related Phenomena in Molecular Materials. *Angew. Chem. Int. Ed.* **2003**, *42*, 268–297. [CrossRef] [PubMed]

20. Ishikawa, N.; Sugita, M.; Ishikawa, T.; Koshihara, S.; Kaizu, Y. Lanthanide Double-Decker Complexes Functioning as Magnets at the Single-Molecular Level. *J. Am. Chem. Soc.* **2003**, *125*, 8694–8695. [CrossRef] [PubMed]

21. Waldmann, O. A Criterion for the Anisotropy Barrier in Single-Molecule Magnets. *Inorg. Chem.* **2007**, *46*, 10035–10037. [CrossRef] [PubMed]

22. Pedersen, K.S.; Woodruff, D.N.; Bendix, J.; Clérac, R. Experimental Aspects of Lanthanide Single-Molecule Magnet Physics. In *Lanthanides and Actinides in Molecular Magnetism*; Wiley-VCH Verlag GmbH & Co. KGaA: Weinheim, Germany, 2015; pp. 125–152.

23. Furrer, A.; Waldmann, O. Magnetic cluster excitations. *Rev. Mod. Phys.* **2013**, *85*, 367–420. [CrossRef]

24. Clemente-Juan, J.M.; Coronado, E.; Gaita-Ariño, A. Mononuclear Lanthanide Complexes: Use of the Crystal Field Theory to Design Single-Ion Magnets and Spin Qubits. In *Lanthanides and Actinides in Molecular Magnetism*; Wiley-VCH Verlag GmbH & Co. KGaA: Weinheim, Germany, 2015; pp. 27–60.

25. Sorace, L.; Gatteschi, D. Electronic Structure and Magnetic Properties of Lanthanide Molecular Complexes. In *Lanthanides and Actinides in Molecular Magnetism*; Wiley-VCH Verlag GmbH & Co. KGaA: Weinheim, Germany, 2015; pp. 1–26.

26. Klokishner, S.I.; Ostrovsky, S.M.; Reu, O.S.; Palii, A.V.; Tregenna-Piggott, P.L.W.; Brock-Nannestad, T.; Bendix, J.; Mutka, H. Magnetic Anisotropy in the $[Cu^{II}LTb^{III}(hfac)_2]_2$ Single Molecule Magnet: Experimental Study and Theoretical Modeling. *J. Phys. Chem. C* **2009**, *113*, 8573–8582. [CrossRef]

27. Magnani, N.; Caciuffo, R.; Colineau, E.; Wastin, F.; Baraldi, A.; Buffagni, E.; Capelletti, R.; Carretta, S.; Mazzera, M.; Adroja, D.T.; et al. Low-energy spectrum of a Tm-based double-decker complex. *Phys. Rev. B* **2009**, *79*, 104407. [CrossRef]

28. Dreiser, J.; Pedersen, K.S.; Piamonteze, C.; Rusponi, S.; Salman, Z.; Ali, M.E.; Schau-Magnussen, M.; Thuesen, C.A.; Piligkos, S.; Weihe, H.; et al. Direct observation of a ferri-to-ferromagnetic transition in a fluoride-bridged 3d-4f molecular cluster. *Chem. Sci.* **2012**, *3*, 1024–1032. [CrossRef]

29. Kofu, M.; Yamamuro, O.; Kajiwara, T.; Yoshimura, Y.; Nakano, M.; Nakajima, K.; Ohira-Kawamura, S.; Kikuchi, T.; Inamura, Y. Hyperfine structure of magnetic excitations in a Tb-based single-molecule magnet studied by high-resolution neutron spectroscopy. *Phys. Rev. B* **2013**, *88*, 64405. [CrossRef]

30. Marx, R.; Moro, F.; Dorfel, M.; Ungur, L.; Waters, M.; Jiang, S.D.; Orlita, M.; Taylor, J.; Frey, W.; Chibotaru, L.F.; et al. Spectroscopic determination of crystal field splittings in lanthanide double deckers. *Chem. Sci.* **2014**, *5*, 3287–3293. [CrossRef]

31. Baker, M.L.; Tanaka, T.; Murakami, R.; Ohira-Kawamura, S.; Nakajima, K.; Ishida, T.; Nojiri, H. Relationship between Torsion and Anisotropic Exchange Coupling in a Tb^{III}-Radical-Based Single-Molecule Magnet. *Inorg. Chem.* **2015**, *54*, 5732–5738. [CrossRef] [PubMed]

32. Pedersen, K.S.; Ungur, L.; Sigrist, M.; Sundt, A.; Schau-Magnussen, M.; Vieru, V.; Mutka, H.; Rols, S.; Weihe, H.; Waldmann, O.; et al. Modifying the properties of 4f single-ion magnets by peripheral ligand functionalisation. *Chem. Sci.* **2014**, *5*, 1650–1660. [CrossRef]

33. Kettles, F.J.; Milway, V.A.; Tuna, F.; Valiente, R.; Thomas, L.H.; Wernsdorfer, W.; Ochsenbein, S.T.; Murrie, M. Exchange Interactions at the Origin of Slow Relaxation of the Magnetization in {TbCu_3} and {DyCu_3} Single-Molecule Magnets. *Inorg. Chem.* **2014**, *53*, 8970–8978. [CrossRef] [PubMed]

34. Giansiracusa, M.J.; Vonci, M.; Van den Heuvel, W.; Gable, R.W.; Moubaraki, B.; Murray, K.S.; Yu, D.; Mole, R.A.; Soncini, A.; Boskovic, C. Carbonate-Bridged Lanthanoid Triangles: Single-Molecule Magnet Behavior, Inelastic Neutron Scattering, and Ab Initio Studies. *Inorg. Chem.* **2016**, *55*, 5201–5214. [CrossRef] [PubMed]

35. Vonci, M.; Giansiracusa, M.J.; Gable, R.W.; Van den Heuvel, W.; Latham, K.; Moubaraki, B.; Murray, K.S.; Yu, D.; Mole, R.A.; Soncini, A.; et al. Ab initio calculations as a quantitative tool in the inelastic neutron scattering study of a single-molecule magnet analogue. *Chem. Commun.* **2016**, *52*, 2091–2094. [CrossRef] [PubMed]

36. Buschow, K.H.J.; de Boer, F.R. *Physics of Magnetism and Magnetic Materials*, 1st ed.; Kluver Academic/Plenum Publishers: New York, NY, USA, 2003.

37. Blagg, R.J.; Ungur, L.; Tuna, F.; Speak, J.; Comar, P.; Collison, D.; Wernsdorfer, W.; McInnes, E.J.L.; Chibotaru, L.F.; Winpenny, R.E.P. Magnetic relaxation pathways in lanthanide single-molecule magnets. *Nat. Chem* **2013**, *5*, 673–678. [CrossRef] [PubMed]

38. Demir, S.; Jeon, I.-R.; Long, J.R.; Harris, T.D. Radical ligand-containing single-molecule magnets. *Coord. Chem. Rev.* **2015**, *289–290*, 149–176. [CrossRef]

39. Bewley, R.I.; Taylor, J.W.; Bennington, S.M. LET, a cold neutron multi-disk chopper spectrometer at ISIS. *Nucl. Instrum. Methods Phys. Res. Sect. A Accel. Spectrom. Detect. Assoc. Equip.* **2011**, *637*, 128–134. [CrossRef]

40. Konstantatos, A.; Bewley, R.; Barra, A.-L.; Bendix, J.; Piligkos, S.; Weihe, H. In-Depth Magnetic Characterization of a [2 × 2] Mn(III) Square Grid Using SQUID Magnetometry, Inelastic Neutron Scattering, and High-Field Electron Paramagnetic Resonance Spectroscopy. *Inorg. Chem.* **2016**, *55*, 10377–10382. [CrossRef] [PubMed]

41. Woolfson, R.J.; Timco, G.A.; Chiesa, A.; Vitorica-Yrezabal, I.J.; Tuna, F.; Guidi, T.; Pavarini, E.; Santini, P.; Carretta, S.; Winpenny, R.E.P. [CrF(O$_2$C^{t}Bu)$_2$]$_9$: Synthesis and Characterization of a Regular Homometallic Ring with an Odd Number of Metal Centers and Electrons. *Angew. Chem. Int. Ed.* **2016**, *55*, 8856–8859. [CrossRef] [PubMed]

42. EPR (and photon-based spectroscopy in general) is subject to the selection rules $\Delta M = \pm 1$ and $\Delta S = 0$, and INS to the rules $\Delta M = 0, \pm 1$ and $\Delta S = 0, \pm 1$. The selection rules of transitions within or between J multiplets are obtained by replacing $M \to M_J, S \to J$.

43. Moreno Pineda, E.; Chilton, N.F.; Marx, R.; Dörfel, M.; Sells, D.O.; Neugebauer, P.; Jiang, S.-D.; Collison, D.; van Slageren, J.; McInnes, E.J.L.; et al. Direct measurement of dysprosium(III)-dysprosium(III) interactions in a single-molecule magnet. *Nat. Commun.* **2014**, *5*, 5243. [CrossRef] [PubMed]

44. Dreiser, J.; Pedersen, K.S.; Schnegg, A.; Holldack, K.; Nehrkorn, J.; Sigrist, M.; Tregenna-Piggott, P.; Mutka, H.; Weihe, H.; Mironov, V.S.; et al. Three-Axis Anisotropic Exchange Coupling in the Single-Molecule Magnets NEt$_4$[Mn^{III}$_2$(5-Brsalen)$_2$(MeOH)$_2$$M^{III}(CN)_6$] ($M$ = Ru, Os). *Chem. Eur. J.* **2013**, *19*, 3693–3701. [CrossRef] [PubMed]

45. Koehler, W.C.; Moon, R.M.; Cable, J.W.; Child, H.R. Neutron scattering experiments on gadolinium. *J. Phys. Colloq.* **1971**, *32*, C1-296–C1-298. [CrossRef]

46. Yu, J.; LeClair, P.R.; Mankey, G.J.; Robertson, J.L.; Crow, M.L.; Tian, W. Exploring the magnetic phase diagram of dysprosium with neutron diffraction. *Phys. Rev. B* **2015**, *91*, 14404. [CrossRef]

47. Sears, V.F. Neutron scattering lengths and cross sections. *Neutron News* **1992**, *3*, 26–37. [CrossRef]

48. Dreiser, J.; Waldmann, O.; Dobe, C.; Carver, G.; Ochsenbein, S.T.; Sieber, A.; Güdel, H.U.; van Duijn, J.; Taylor, J.; Podlesnyak, A. Quantized antiferromagnetic spin waves in the molecular Heisenberg ring CsFe$_8$. *Phys. Rev. B* **2010**, *81*, 24408. [CrossRef]

49. Ungur, L.; Chibotaru, L.F. Computational Modelling of the Magnetic Properties of Lanthanide Compounds. In *Lanthanides and Actinides in Molecular Magnetism*; Wiley-VCH Verlag GmbH & Co. KGaA: Weinheim, Germany, 2015; pp. 153–184.

50. Furrer, A.; Güdel, H.U.; Blank, H.; Heidemann, A. Direct Observation of Exchange Splittings in Cs$_3$Tb$_2$Br$_9$ by Neutron Spectroscopy. *Phys. Rev. Lett.* **1989**, *62*, 210–213. [CrossRef] [PubMed]

51. Furrer, A.; Güdel, H.U.; Krausz, E.R.; Blank, H. Neutron spectroscopic study of anisotropic exchange in the dimer compound Cs$_3$Ho$_2$Br$_9$. *Phys. Rev. Lett.* **1990**, *64*, 68–71. [CrossRef] [PubMed]

52. Guo, Y.-N.; Ungur, L.; Granroth, G.E.; Powell, A.K.; Wu, C.; Nagler, S.E.; Tang, J.; Chibotaru, L.F.; Cui, D. An NCN-pincer ligand dysprosium single-ion magnet showing magnetic relaxation via the second excited state. *Sci. Rep.* **2014**, *4*, 5471. [CrossRef] [PubMed]

53. Hutchings, M.T. Point-Charge Calculations of Energy Levels of Magnetic Ions in Crystalline Electric Fields. In *Solid State Physics*; Seitz, F., Turnbull, D., Eds.; Elsevier B.V.: Amsterdam, The Netherlands, 1964; Volume 16, pp. 227–273.

54. Baker, M.L.; Guidi, T.; Carretta, S.; Ollivier, J.; Mutka, H.; Gudel, H.U.; Timco, G.A.; McInnes, E.J.L.; Amoretti, G.; Winpenny, R.E.P.; et al. Spin dynamics of molecular nanomagnets unravelled at atomic scale by four-dimensional inelastic neutron scattering. *Nat. Phys.* **2012**, *8*, 906–911. [CrossRef]

55. Guidi, T.; Gillon, B.; Mason, S.A.; Garlatti, E.; Carretta, S.; Santini, P.; Stunault, A.; Caciuffo, R.; van Slageren, J.; Klemke, B.; et al. Direct observation of finite size effects in chains of antiferromagnetically coupled spins. *Nat. Commun.* **2015**, *6*, 7061. [CrossRef] [PubMed]

56. Waldmann, O.; Bircher, R.; Carver, G.; Sieber, A.; Güdel, H.U.; Mutka, H. Exchange-coupling constants, spin density map, and Q dependence of the inelastic neutron scattering intensity in single-molecule magnets. *Phys. Rev. B* **2007**, *75*, 174438. [CrossRef]

57. Borta, A.; Gillon, B.; Gukasov, A.; Cousson, A.; Luneau, D.; Jeanneau, E.; Ciumacov, I.; Sakiyama, H.; Tone, K.; Mikuriya, M. Local magnetic moments in a dinuclear Co^{2+} complex as seen by polarized neutron diffraction: Beyond the effective spin-1/2 model. *Phys. Rev. B* **2011**, *83*, 184429. [CrossRef]

58. Zaharko, O.; Pregelj, M.; Zorko, A.; Podgajny, R.; Gukasov, A.; van Tol, J.; Klokishner, S.I.; Ostrovsky, S.; Delley, B. Source of magnetic anisotropy in quasi-two-dimensional XY {Cu$_4$(tetrenH$_5$)W(CN)$_8$]$_4$·7.2 H$_2$O}$_n$ bilayer molecular magnet. *Phys. Rev. B* **2013**, *87*, 24406. [CrossRef]

59. Ridier, K.; Gillon, B.; Gukasov, A.; Chaboussant, G.; Cousson, A.; Luneau, D.; Borta, A.; Jacquot, J.-F.; Checa, R.; Chiba, Y.; et al. Polarized Neutron Diffraction as a Tool for Mapping Molecular Magnetic Anisotropy: Local Susceptibility Tensors in CoII Complexes. *Chem. Eur. J.* **2016**, *22*, 724–735. [CrossRef] [PubMed]

60. Ridier, K.; Mondal, A.; Boilleau, C.; Cador, O.; Gillon, B.; Chaboussant, G.; Le Guennic, B.; Costuas, K.; Lescouëzec, R. Polarized Neutron Diffraction to Probe Local Magnetic Anisotropy of a Low-Spin Fe(III) Complex. *Angew. Chem. Int. Ed.* **2016**, *55*, 3963–3967. [CrossRef] [PubMed]

61. Evans, W.J.; Lee, D.S.; Rego, D.B.; Perotti, J.M.; Kozimor, S.A.; Moore, E.K.; Ziller, J.W. Expanding Dinitrogen Reduction Chemistry to Trivalent Lanthanides via the LnZ$_3$/Alkali Metal Reduction System: Evaluation of the Generality of Forming Ln$_2$(μ-η^2:η^2-N$_2$) Complexes via LnZ$_3$/K. *J. Am. Chem. Soc.* **2004**, *126*, 14574–14582. [CrossRef] [PubMed]

62. Schnack, J. Influence of intermolecular interactions on magnetic observables. *Phys. Rev. B* **2016**, *93*, 54421. [CrossRef]

63. Albuquerque, A.F.; Alet, F.; Corboz, P.; Dayal, P.; Feiguin, A.; Fuchs, S.; Gamper, L.; Gull, E.; Gürtler, S.; Honecker, A.; et al. The ALPS project release 1.3: Open-source software for strongly correlated systems. *J. Magn. Magn. Mater.* **2007**, *310*, 1187–1193. [CrossRef]

64. Bauer, B.; Carr, L.D.; Evertz, L.D.; Feiguin, A.; Freire, J.; Fuchs, S.; Gamper, L.; Gukelberger, J.; Gull, E.; Guertler, S.; et al. The ALPS project release 2.0: Open source software for strongly correlated systems. *J. Stat. Mech. Theory Exp.* **2011**, *2011*, P05001. [CrossRef]

65. Bain, G.A.; Berry, J.F. Diamagnetic Corrections and Pascal's Constants. *J. Chem. Educ.* **2008**, *85*, 532. [CrossRef]

magnetochemistry

MDPI

Article

Hybrid Molecular Compound Exhibiting Slow Magnetic Relaxation and Electrical Conductivity

Yongbing Shen [1], Goulven Cosquer [1,2], Brian K. Breedlove [1] and Masahiro Yamashita [1,2,3,*]

[1] Department of Chemistry, Graduate School of Science, Tohoku University, Aramaki-Aza-Aoba, Aoba-ku, Sendai 980-8578, Japan; shenyongbing17@gmail.com (Y.S.); cosquer.g@m.tohoku.ac.jp (G.C.); breedlove@m.tohoku.ac.jp (B.K.B.)
[2] Core Research for Evolutional Science and Technology (CREST), Japan Science and Technology (JST), 4-1-8 Kawaguchi, Saitama 332-0012, Japan
[3] WPI Research Center, Advanced Institute for Materials Research, Tohoku University, 2-1-1 Katahira, Aoba-ku, Sendai 980-8577, Japan
[*] Correspondence: yamashita.m@gmail.com; Tel.: +81-22-765-6547

Academic Editor: Kevin Bernot
Received: 27 September 2016; Accepted: 30 November 2016; Published: 9 December 2016

Abstract: Electrochemical oxidation of a solution containing $KDy(hfac)_4$ (hfac, hexafluoroacetyacetone) and Bis(ethylenedithio)tetrathiafulvalene (BEDT-TTF) afforded a hybrid material formulated as $[\beta'-(BEDT-TTF)_2Dy(CF_3COO)_4 \cdot MeCN]_n$. The complex crystallizes in the triclinic space group $P\bar{1}$. The before mentioned complex has a chain structure containing $4f$ ions bridged by mono-anion CF_3COO^- ligand, and acts as single-molecule magnet (SMM) at low temperature. The conducting layer was composed of partially oxidized BEDT-TTF molecules in β' type arrangement. The presence of radical cation and its charge ordering was assigned on the basis of optical spectra. Electrical resistivity measurements revealed semiconducting behaviour (conductivity at room temperature of 1.1×10^{-3} S·cm^{-1}, activation energy of 158.5 meV) at ambient pressure.

Keywords: BEDT-TTF; conductivity; SMM; dysprosium

1. Introduction

Hybrid molecular materials combining conductivity (delocalized electrons or holes) and magnetism (localized electrons) have been intensively studied in the past decades, in order to observe a synergy between these properties [1–5]. Organic conductors, such as bis(ethylenedithio)tetrathiafulvalene (BEDT-TTF), and M(dmit)$^-$ (M: 3d or 4d metal; dmit: 4,5-dimercapto-1,3-dithiole-2-thione) with π electrons have been widely used in conducting materials, affording a large number of superconductors [3], such as paramagnet/superconductor, anti-ferromagnet/superconductor, and ferromagnet/metal [6–8]. In recent years, research has been performed using a single-molecule magnet (SMM) as an electronic conductor or a valve. The molecule is placed between two gold electrodes, and it acts as an electron transport. In the case of polarized spin, the SMM acts as a valve in relation to its magnetic polarization [9].

In parallel, several groups have synthesized materials combining SMM behaviour and molecular conductivity [10–13]. SMMs are isolated molecules possessing individual large ground state spins and uniaxial anisotropies, which cause a finite energy barrier (Δ) between up and down spin states. SMMs are characterized by slow relaxation of magnetization and quantum phenomena, such as quantum tunnelling of magnetization (QTM), which can be used to design spintronics devices. However, since both SMM behaviour and superconductivity occur at low temperature (around 15 K or below), there is a possibility that both can occur at the same time.

Our work focused on this third strategy. In previously reported materials, SMMs acted as donors, and organic conductors acted as acceptors leading to coexistence of SMM behaviour and semi-conductivity but in a different temperature range. Here, we reversed the roles of the donor and acceptor. We report a hybrid material, $[\beta'\text{-}(BEDT\text{-}TTF)_2Dy(CF_3COO)_4\cdot MeCN]_n$ (**1**), with an anionic Dy^{III} complex exhibiting slow relaxation of magnetisation and the organic conductor BEDT-TTF. To the best of our knowledge, this is one of few works in which slow relaxation of the magnetization from an f ion and molecular semi-conductivity have been combined [14].

2. Results

The electrochemical oxidation of BEDT-TTF and cocrystallisation with $Dy(hfac)_4$ (hfac, hexafluoroacetyacetone) afforded the polymeric complex (**1**). The degradation of $Dy(hfac)_4$ to $\{Dy(CF_3COO)_4\}_n$ occurred when a non-dry solvent was used. The water in the solvent along with the catalytic action of the dc current decomposed the $hfac^-$ ligand to CF_3COCH_3 and CF_3COO^- [15].

2.1. Crystal Structure

Compound (**1**) crystallised in the triclinic $P\bar{1}$ space group (Table S1, Figure S1) with two BEDT-TTF units, one $[Dy(CF_3COO)_4]^{-1}$ unit and one acetonitrile molecule in the asymmetric unit (Figure 1). The crystal packing had alternating organic and inorganic layers along the c axis (Figure 2). The inorganic layer contains "zig-zag" chains of dysprosium ions bridged by deprotonated trifluoroacetic acid ligands to form a paddle wheel infinite chain. The coordination sphere of the dysprosium ions has a quasi-perfect D_{4d} symmetry (deviation of 0.029 obtained by SHAPE [16,17]) composed of eight oxygen atoms (Table S2). The distance between adjacent dysprosium metals was determined to be 4.424 Å and 4.548 Å with a zig-zag angle of 151.83°. The chains are stacked parallel along the b axis with an inter-chain distance of 11.541 Å. The inorganic layers were separated by 24.05 Å along the c axis. The organic layer contains BEDT-TTF molecules in a β'-phase [18–20]. In this layer (Figure 3), the short S$\cdots$S distance helped to form a ladder-like motif with distances in the range of 3.341–3.643 Å, which are shorter than the sum of the Van der Waals radii (3.7 Å). The charges of each BEDT-TTF molecule were estimated to be +0.22 and +0.59 [21]. This total charge of +0.81 is smaller than the theoretical charge of +1 imposed by the electroneutrality of the complex. The charge was determined only on the basis of the distance between the atoms of the BEDT-TTF core. Interactions between the molecules deform the core (elongation or compression of the bonds), which affects the estimation of the charge, explaining the difference between the estimated and theoretical charges. This estimation is just a tool to describe the charge repartition in the material and therefore, it may not be accurate. Moreover, X-ray diffraction affords an averaged structure, meaning that the charge is not estimated from a single molecule but from an averaged molecule. This averaging explains the non-integer charge of each BEDT-TTF.

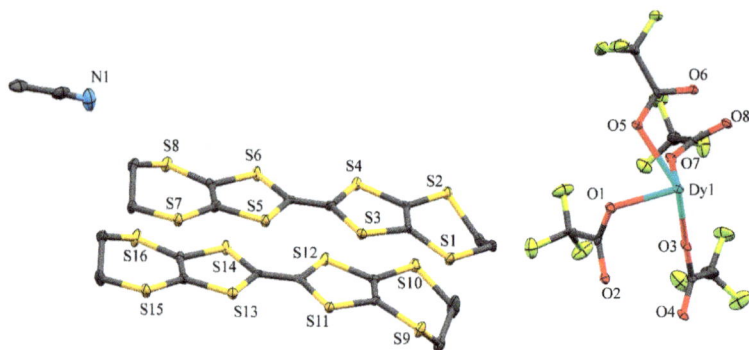

Figure 1. Asymmetric unit of [β′-(BEDT-TTF)$_2$Dy(CF$_3$COO)$_4$·MeCN]$_n$ (**1**) with hydrogen atoms omitted for clarity.

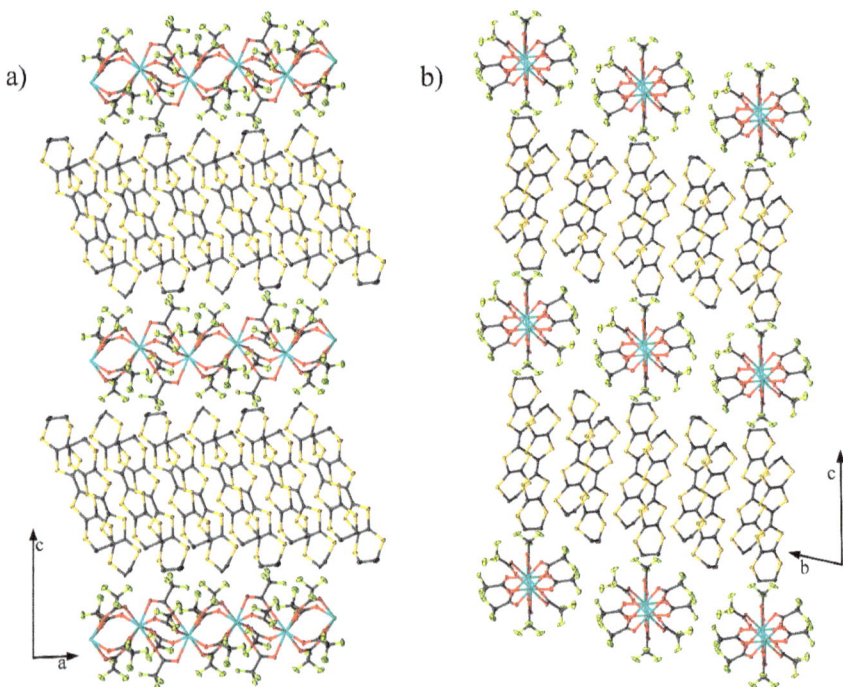

Figure 2. Packing structure in (**a**) the *ac* and (**b**) *bc* planes. Solvent molecule and hydrogen atoms were omitted for clarity.

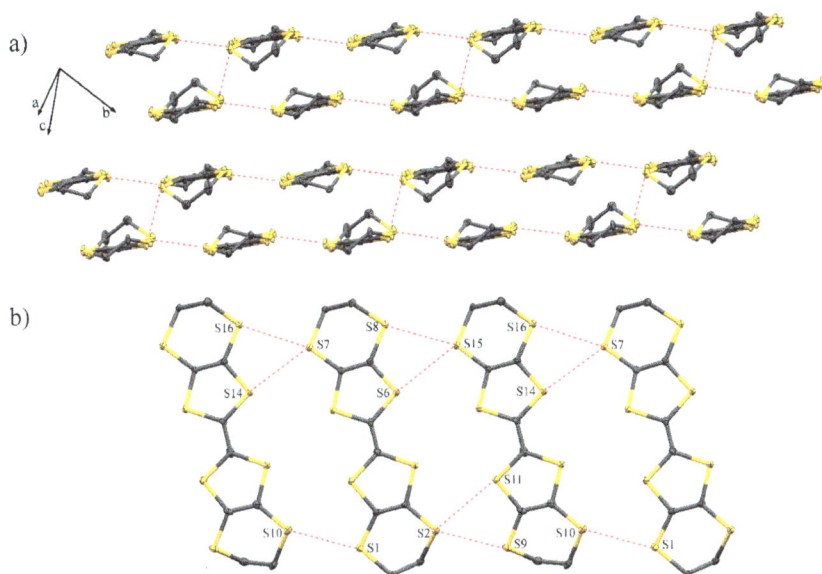

Figure 3. (a) View of the β'-phase packing where the short S···S intermolecular contacts are highlighted by red dashed lines and (b) details of one of the sheets.

2.2. Optical Properties

In the UV-Vis spectra of (**1**), a strong absorption peak was observed centred at 980 nm (Figure S2), which was attributed to the electron transition from SOMO–1 to SOMO of BEDT-TTF$^{+\bullet}$. This absorption peak is evidence for a radical in (**1**), which agrees with the total charge of 0.81 electrons estimated from the crystal structure. In the higher energy region, the absorption peak at 461 nm was ascribed to a π–π^* transition of BEDT-TTF$^{+\bullet}$ [22,23].

In order to investigate the electronic structure of (**1**), polarized IR reflectance spectra were acquired at 300 K. The spectra were polarized along the BEDT-TTF stacking direction, and the electrical vector was parallel to the [110] direction (Figure 4). The broad peak around 3500 cm^{-1} was attributed to an intermolecular charge transfer (CT) between two (BEDT-TTF)$^{\bullet+}$ moieties. In the 900–1800 cm^{-1} region, the three peaks (1672 cm^{-1}, 1349 cm^{-1} and 1270 cm^{-1}) were attributed to the ν_{27} stretching mode of neutral BEDT-TTF molecule and the ν_3 stretching mode of the radical BEDT-TTF molecule. The broad maximum around 1300 cm^{-1} was attributed to an electron-molecular vibrational (e-mv) interaction due to coupling between intermolecular CT and C=C stretching modes.

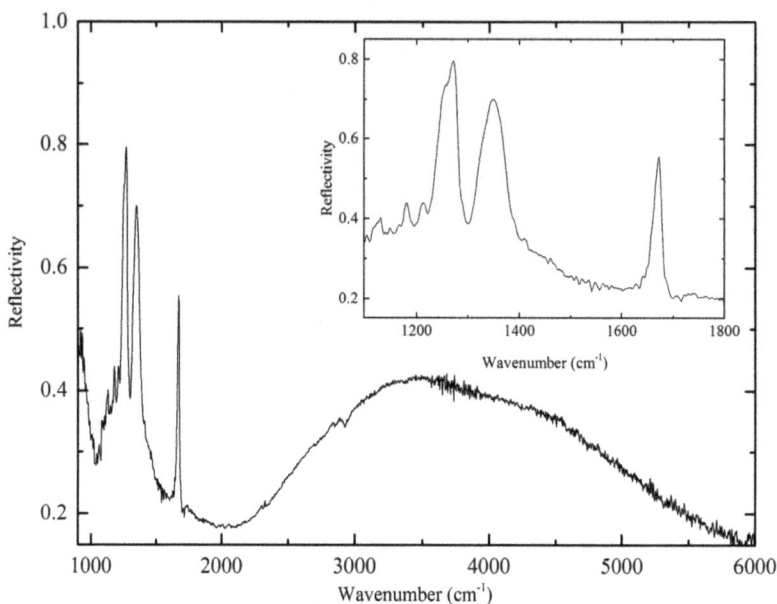

Figure 4. Polarized IR reflectance spectrum at 300 K.

The electronic state of the BEDT-TTF molecules were estimated from the position and intensity of vibration peaks [24,25]. The charges of the BEDT-TTF molecules obtained from X-ray crystal structure analyses and the polarized IR measurement agree with each other with charge ordering of one neutral BEDT-TTF molecule and one radical cation BEDT-TTF molecule along the [110] direction. The charge-ordered state is supported by the electrical resistivity [26–31].

2.3. Electrical Conductivity

The electronic conductivity of the single-crystal was measured using a two-probe method in the temperature range of 300–100 K (Figure S3). Below 100 K, the conductivity of the crystals was out of the range for our equipment. In (**1**), semiconductor behaviour was shown along the *b* axis. The conductivity at room temperature (σ_{rt}) was 1.7×10^{-3} S·cm^{-1} and decreased gradually with a decrease in temperature. The activation energy (E_a) between the valence and conduction bands was calculated to be 158.5 meV at ambient pressure. Magneto-resistance has been investigated but not clearly observed due to the high resistance of the sample and/or limitation of our equipment.

Applying isostatic pressure in the range 0.4–2.2 GPa induced a continuous increase of three orders of magnitude in the conductivity at 100 K (Figure 5). E_a decreased gradually with an increase in the pressure. This decrease in E_a indicates an enhancement in the conduction band [32], which is attributed to isostatic compression of the BEDT molecule and an increase in the overlap of the molecular orbitals. Another possibility is a structural re-arrangement of the molecule, especially a change in the dihedral angle between neighbouring molecules which causes an overlap in the orbitals, which reduces E_a [33].

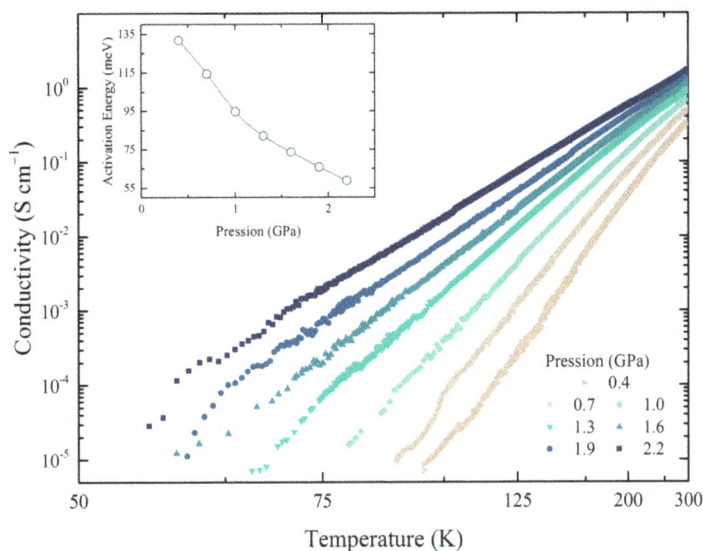

Figure 5. Temperature and pressure dependence of the conductivity. In the insert, the activation energy versus pressure.

2.4. Magnetic Properties

The temperature dependence of the static magnetic susceptibility (χ) was measured on a polycrystalline sample by applying a field of 1000 Oe (Figure 6). The χT value of 13.30 cm^3·K·mol^{-1} at 300 K was significantly lower than the expected value of 14.545 cm^3·K·mol^{-1} for non-interacting radicals and free Dy^{3+} ions. This difference can be attributed to a preferencial orientation of the crystallite, which induce a deviation from the expected isotrope value [34]. χT gradually decreased when the temperature was decreased, reaching a minimum value of 10.15 cm^3·K·mol^{-1} at 7.5 K. Then it increased to a value of 10.94 cm^3·K·mol^{-1} at 2 K. This increase below 7.5 K was attributed to dipole-dipole interactions between BEDT-TTF cation radicals and/or Dy ions [35]. Analysis of the ln(χT) versus $1/T$ plots shows two linear regimes in the range of 7–5.5 K and 5–2 K, which were fit with the equation $\chi T = C_{\text{eff}} \times e^{\Delta/kT}$ (Figure S4). Nevertheless, the values of Δ are very small and are not clear evidence of single-chain magnet (SCM) behaviour for this complex. Interaction(s) such as dipole-dipole interaction definitely occur between the ions that exist in this compound but they are not strong enough to cause long-range ordering of the magnetic moment, characteristic of SCMs [36].

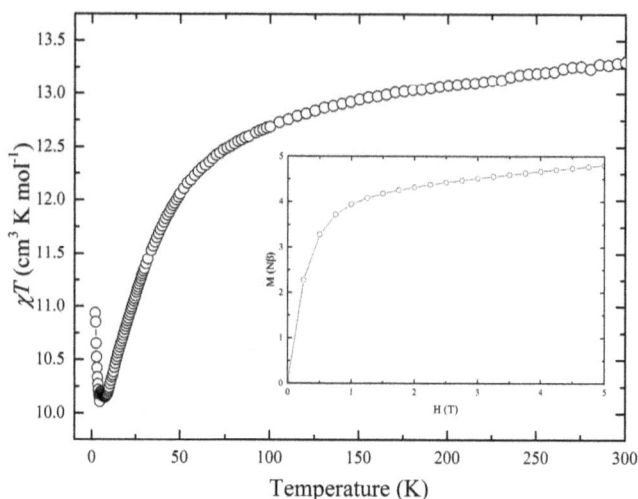

Figure 6. Temperature dependence of χT for a polycrystalline sample. Insert shows the magnetisation curve at 1.85 K.

The magnetisation curve exhibited pseudo-saturation from 1.5 T with a linear slope of 0.16 N·β^{-1}, reaching a value of 4.81 N·β at 5 T (insert of Figure 6). No hysteresis was observed.

The dynamic susceptibility exhibited weak temperature and frequency dependences below 5 K (Figures S5–S7 and Table S3). However, distinct out-of-phase peaks were not observed over the full temperature range up to 1000 Hz due to the merging of two relaxation times. The data was analysed by using dual relaxation Cole-Cole model with one of the peaks over 1000 Hz and the adiabatic susceptibility set to zero [37]. The faster process, which was completely out of range, was used only to allow us to determine the nature of the slower process: a combination of Orbach and quantum tunnelling of the magnetisation (QTM) (Table 1). By applying an external magnetic field, it was possible to partially unmerge the two peaks (Figure S8) with an optimal field at 1000 Oe (Figures 7 and S9 and Table S4). With and without external field, the second peak was out of the range of our equipment and could not be determined accurately enough to be discussed here. The field suppresses the QTM and allows the system to relax though a direct and Orbach process (Figure 8). As expected, the energy barrier and the pre-exponential factor are comparable with and without a field and are comparable with other Dy^{III} SMMs [37]. A Raman relaxation process has been considered to be a possible mechanism. However, it does not match with the experimental data. The origin of the two peaks is still unclear, and further studies are needed. Preparation of a diamagnetic doped compound has been tried to investigate the role of the dipole interactions in the relaxation process. However, it has been unsuccessful so far.

Table 1. Detail information about magnetic properties.

Field	0 Oe	1000 Oe (Low f)
Calculation Equation	$\tau^{-1} = \frac{1}{\tau_0}\exp\left(\frac{-\Delta}{k_B T}\right) + \text{QTM}$	$\tau^{-1} = \frac{1}{\tau_0}\exp\left(\frac{-\Delta}{k_B T}\right) + AH^4T^2$
A	-	3.3×10^{-12}
τ0 (s)	8.8×10^{-8}	5.0×10^{-8}
Δ (cm^{-1})	21.3	22.1
QTM (s)	1.1×10^{-3}	-

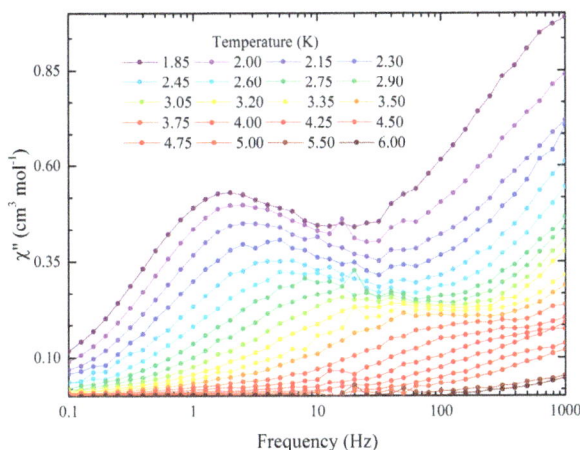

Figure 7. Frequency dependence of the out-of-phase magnetic susceptibility in a 1000 Oe field as a function of temperature.

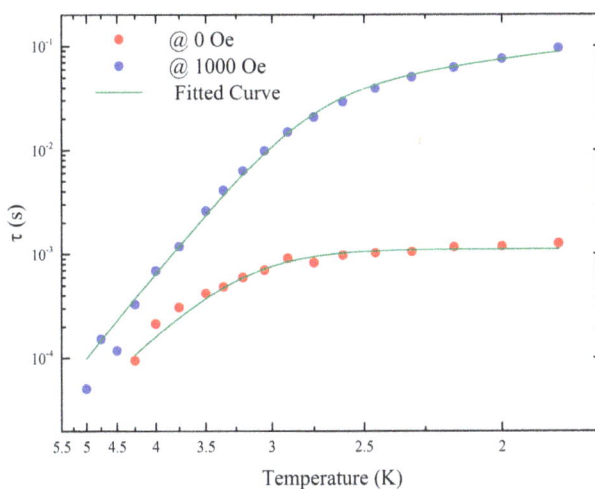

Figure 8. Temperature dependence of relaxation time in 0 and 1000 Oe dc fields.

3. Materials and Methods

3.1. Synthesis

Solvents were used without further purification. KDy(hfac)$_4$ was prepared following the reported methods [38]. BEDT-TTF was purchased from TCI Tokyo Chemical Industry Co., LTD, Tokyo, Japan.

[β′-(BEDT-TTF)$_2$Dy(CF$_3$COO)$_4$·MeCN]$_n$ (**1**). Single crystals of (**1**) were synthesized by using electrochemical oxidation. To an acetonitrile solution (10 mL) of KDy(hfac)$_4$ (100 mg, 0.1 mmol) and 18-crown-6-ether (100 mg, 0.38 mmol), a dichloromethane solution (15 mL) of BEDT-TTF (12 mg, 0.03 mmol) was slowly added. The mixture was stirred for 1 h to produce an orange brown solution. Black plate crystals suitable for X-ray analysis (CCDC 1506816) and resistivity measurements were obtained at the anode after few days by applying a constant current of 0.7 μA. Crystals were washed with methanol and dried in the air (yield: 7.84 mg 5.7 μmol, 5.7%). IR (cm^{-1}): 1770, 1665, 1457, 1326, 1290, 1209.

3.2. Physical Measurements

UV-Vis spectra were acquired for solid-state samples, using a KBr disk, on a Shimadzu UV-3100pc (Shimadzu, Kyoto, Japan). Reflectance IR spectra were acquired on a JASCO IRT-5000 microscope and a FT-IR-6200YMS Infrared spectrometer (JASCO, Tokyo, Japan) in the ac plane of a single crystal with an angle of 60° with the surface.

Magnetic susceptibility measurements were conducted using a Quantum Design SQUID magnetometer MPMS-5L (Quantum Design, San Diego, CA, USA) in the temperature and dc field ranges of 1.8–300 K and −5–5 T, respectively. AC measurements were performed at frequencies in the range of 0.1–1000 Hz with an ac field amplitude of 3 Oe. A polycrystalline sample embedded in n-eicosane was used for the measurements.

The temperature dependence of the electrical resistivity was measured using a Quantum Design PPMS 6000 (Quantum Design, San Diego, CA, USA) with an external Keithley 2611 System SourceMeter (Keithley Instruments, Solon, OH, USA) by using a two-probe method at ambient pressure. Gold wires (15 μm diameter) were attached to the crystal with carbon paste.

The electrical resistivity under pressure was measured with a Be-Cu clamp-type cell using a four-probe method. Pressure was applied via Daphne 7373 oil at room temperature and clamped with screws. The pressure value decreases by 0.2 GPa at low temperatures compared to that at room temperature.

Single-crystal crystallographic data were collected at 103 K on a Rigaku Saturn70 CCD Diffractometer (Rigaku, Tokyo, Japan) with graphite-monochromated Mo Kα radiation (λ = 0.71075 Å) produced using a VariMax microfocus X-ray rotating anode source. A single crystal with dimensions of 0.20 × 0.07 × 0.01 mm^3 was used. Data processing was performed using the Crystal Clear crystallographic software package [39]. The structures were solved by using direct methods using SIR-92 [40]. Refinement was carried out using WinGX 2013.3 package [41] and SHELXL-2013 [42]. The non-H atoms were refined anisotropically using weighted full-matrix least squares on F. H atoms attached to the C atoms were positioned using idealized geometries and refined using a riding model.

Powder X-ray diffraction was performed on a Bruker AXS D2 phaser (Bruker Corporation Billerica, MA, USA).

4. Conclusions

Hybrid material (**1**) composed of an anionic magnetic chain layer and 2D cationic conducting layer (with the anionic/cationic configuration opposite to that previously reported), was prepared. The presence of a radical, centred on the BEDT-TTF, was evidenced by a strong SOMO→SOMO−1 transition in the UV-Vis spectra and by the electron-molecular vibrational interactions observed by using polarized IR spectroscopy. The conductivity at room temperature was determined to be 1.7 × 10^{-3} S·cm^{-1} with an activation energy of 158.5 meV. Slow relaxation of the magnetisation was clearly observed in a 1000 Oe without any hysteresis. No correlation between conductivity and magnetism was observed due to the difference in temperature ranges for each property. This preliminary study has provided us with significant information for designing new hybrid materials based on quantum magnets and molecular conductors by increasing the molecular interactions between them. In the future, particular attention will be paid to the substitution of the trifluoro-methyl group.

Supplementary Materials: The following are available online at www.mdpi.com/2312-7481/2/4/44/s1, Figure S1: Powder X-ray diffraction spectra for **1**, Figure S2: UV-Vis Spectra, Figure S3: Resistivity curve, Figure S4: Plots of ln(χT) versus 1/T, Figure S5: Temperature dependence of ac susceptibility measured in 0 Oe dc fields, Figure S6: Out-of-phase signal of the susceptibility of complex **1** without external field, Figure S7: Normalized Argand plot for complex **1** without external field, Figure S8: Frequency dependence of χ" at 1.85 K as a function of dc field, Figure S9: Normalized Argand plot for complex **1** in a 1000 Oe external field, Table S1: Crystallographic data for **1**, Table S2: Coordination geometry deviation, Table S3: Fitting parameter of frequency dependence of susceptibility for **1** in 0 Oe field, Table S4: Fitting parameters for frequency dependence of susceptibility for **1** in a 1000 Oe field.

Acknowledgments: This work was financially supported by Core Research for Evolutional Science and Technology (CREST), Japan Science and Technology (JST). Yoji Horii and Takefumi Yoshida are acknowledged for their help with the solid-state UV-Vis and polarized IR measurements. In addition, we express our acknowledgement to Hiroshi Ito from the Department of Applied Physics, Nagoya University, Nagoya, Japan and his co-worker for the measurement of the pressure and temperature dependence of the conductivity.

Author Contributions: M.Y. conceived and designed the experiments; Y.S. performed the experiments; Y.S. and G.C. analyzed the data; Y.S., G.C. and B.K.B. wrote the paper.

Conflicts of Interest: The authors declare no conflict of interest. The founding sponsors had no role in the design of the study; in the collection, analyses, or interpretation of data; in the writing of the manuscript, and in the decision to publish the results.

References

1. Ouahab, L. Organic/inorganic supramolecular assemblies and synergy between physical properties. *Chem. Mater.* **1997**, *9*, 1909–1926. [CrossRef]
2. Enoki, T.; Miyazaki, A. Magnetic TTF-based charge-transfer complexes. *Chem. Rev.* **2004**, *104*, 5449–5478. [CrossRef] [PubMed]
3. Williams, J.M.; Ferraro, J.R.; Thorn, R.J.; Carlson, K.D.; Geiger, U.; Wang, H.H.; Kini, A.M.; Whangbo, M.H. *Organic Superconductors: Synthesis, Structure, Properties and Theory*; Grimes, R.N., Ed.; Prentice Hall: Englewood Cliffs, NJ, USA, 1992.
4. Martin, L.; Turner, S.S.; Day, P.; Mabbs, F.E.; McInnes, E.J.L. New molecular superconductor containing paramagnetic chromium (iii) ions. *Chem. Commun.* **1997**, *15*, 1367–1368. [CrossRef]
5. Fujiwara, H.; Fujiwara, E.; Nakazawa, Y.; Narymbetov, B.Z.; Kato, K.; Kobayashi, H.; Kobayashi, A.; Tokumoto, M.; Cassoux, P. A novel antiferromagnetic organic superconductor *k*-(BETS)$_2$FeBr$_4$ [Where BETS = Bis(ethylenedithio) tetraselenafulvalene]. *J. Am. Chem. Soc.* **2001**, *123*, 306–314. [CrossRef] [PubMed]
6. Yamaguchi, K.; Kitagawa, Y.; Onishi, T.; Isobe, H.; Kawakami, T.; Nagao, H.; Takamizawa, S. Spin-mediated superconductivity in cuprates, organic conductors and π–d conjugated systems. *Coord. Chem. Rev.* **2002**, *226*, 235–249. [CrossRef]
7. Day, P.; Kurmoo, M.; Mallah, T.; Marsden, I.R.; Friend, R.H.; Pratt, F.L.; Hayes, W.; Chasseau, D.; Gaultier, J.; Bravic, G.; et al. Structure and properties of tris[bis(ethylenedithio) tetrathiaful-valenium]tetrachlorocopper hydrate (BEDT-TTF)$_3$CuCl$_4$·H$_2$O: First evidence for coexistence of localized and conduction electrons in a metallic charge-transfer salt. *J. Am. Chem. Soc.* **1992**, *114*, 10722–10729. [CrossRef]
8. Ojima, E.; Fujiwara, H.; Kato, K.; Kobayashi, H.; Tanaka, H.; Kobayashi, A.; Tokumoto, M.; Cassoux, P. Antiferromagnetic Organic Metal Exhibiting Superconducting Transition, K-(BETS)2FeBr4 [BETS Bis(ethylenedithio) tetraselenafulvalene]. *J. Am. Chem. Soc.* **1999**, *121*, 5581–5582. [CrossRef]
9. Bogani, L.; Wernsdorfer, W. Molecular spintronics using single-molecule magnets. *Nat. Mater.* **2008**, *7*, 179–186. [CrossRef] [PubMed]
10. Kurmoo, M.; Graham, A.W.; Day, P.; Coles, S.J.; Hursthouse, M.B.; Caulfield, J.L.; Singleton, J.; Pratt, F.L.; Hayes, W.; Ducasse, L.; et al. Superconducting and Semiconducting Magnetic Charge Transfer Salt: (BEDT-TTF)$_4$AFe(C$_2$O$_4$)$_3$C$_6$H$_5$CN (A = H$_2$O, K, NH$_4$). *J. Am. Chem. Soc.* **1995**, *117*, 12209–12227. [CrossRef]
11. Hiraga, H.; Miyasaka, H.; Nakata, K.; Kajiwara, K.; Takaishi, S.; Oshima, Y.; Nojiri, H.; Yamashita, M. Hybrid molecular materials exhibiting single-molecule magnet behaviour and molecular conductivity. *Inorg. Chem.* **2007**, *46*, 9661–9671. [CrossRef] [PubMed]
12. Hiraga, H.; Miyasaka, H.; Takaishi, S.; Kajiwara, T.; Yamashita, M. Hybridized complexes of $\frac{1}{2}$ MnIII 2 single-molecule magnets and Ni dithiolate complexes. *Inorg. Chim. Acta* **2008**, *361*, 3863–3872. [CrossRef]
13. Kubo, K.; Shiga, T.; Yamamoto, T.; Tajima, A.; Moriwaki, T.; Ikemoto, Y.; Yamashita, M.; Sessini, E.; Mercuri, M.-L.; Deplano, P.; et al. Electronic state of a conducting single molecule magnet based on Mn-salen type and Ni-Dithiolene complexes. *Inorg. Chem.* **2011**, *50*, 9337–9344. [CrossRef] [PubMed]
14. Ueki, S.; Nogami, T.; Ishida, T.; Tamura, M. ET and TTF salts with lanthanide complex ions showing frequency-dependent ac magnetic susceptibility. *Mol. Cryst. Liq. Cryst.* **2006**, *455*, 129–134. [CrossRef]
15. Johnson, D.A.; Waugh, A.B.; Hambley, T.W.; Taylor, J.C. Synthesis and Crystal Structure of 1,1,1,5,5,5-Hexafluoro-2-aminopentan-4-one (HFAP). *J. Fluor. Chem.* **1985**, *27*, 371–378. [CrossRef]

16. Ruiz-Martínez, A.; Casanova, D.; Alvarez, S. Polyhedral structures with an odd number of vertices: Nine-coordinate metal compounds. *Chem. Eur. J.* **2008**, *14*, 1291–1303. [CrossRef] [PubMed]

17. Alvarez, S.; Alemany, P.; Casanova, D.; Cirera, J.; Llunell, M.; Avnir, D. Shape maps and polyhedral interconversion paths in transition metal chemistry. *Coord. Chem. Rev.* **2005**, *249*, 1693–1708.

18. Mori, T. Structural genealogy of BEDT-TTF-based organic conductors I. parallel molecules: β and β Phases. *Bull. Chem. Soc. Jpn.* **1998**, *71*, 2509–2526. [CrossRef]

19. Mori, T.; Mori, H.; Tanaka, S. Structural genealogy of BEDT-TTF-based organic conductors II. inclined molecules: θ, α, and κ Phases. *Bull. Chem. Soc. Jpn.* **1999**, *72*, 179–197. [CrossRef]

20. Shibaeva, R.P.; Yagubskii, E.B. Molecular conductors and superconductors based on Trihalides of BEDT-TTF and some of its analogues. *Chem. Rev.* **2004**, *104*, 5347–5378. [CrossRef] [PubMed]

21. Guionneau, P.; Kepert, C.J.; Bravic, G.; Chasseau, D.; Truter, M.R.; Kurmoo, M.; Day, P. Determining the charge distribution in BEDT-TTF salts. *Synth. Metal.* **1997**, *86*, 1973–1974. [CrossRef]

22. Rosokha, S.V.; Kochi, J.K. Molecule and electronic structure of the long-bonded π-dimers of tetrathiafulvalene cation-radical in intermolecular electron transfer and in (solid-state) conductivity. *J. Am. Chem. Soc.* **2007**, *129*, 828–838. [CrossRef] [PubMed]

23. Cosquer, G.; Pointillart, F.; Le Guennic, B.; Le Gal, Y.; Golhen, S.; Cador, O.; Ouahab, L. 3d4f heterobimetallic dinuclear and tetranuclear complexes. *Inorg. Chem.* **2012**, *51*, 8488–8501. [CrossRef] [PubMed]

24. Pokhodnya, K.I.; Cassoux, P.; Feltre, L.; Meneghtti, M. Optical excitations in a quarter-filled Ni(dmit)$_2$ based compound described by a dimerized octamer model. *Synth. Met.* **1999**, *103*, 2187. [CrossRef]

25. Romaniello, P.; Lelj, F.; Arca, M.; Devillanova, F.A. Structural and new spectroscopic properties of neutral [M(dmit)$_2$](dmit = C$_3$S$_5$$^{2-}$, 1,3-dithiole-2-thione-4,5-dithiolate) and [M(H$_2$timdt)$_2$](H$_2$timdt = H$_2$C$_3$N$_2$S$_3$$^{1-}$, monoanion of imidazolidine-2,4,5-trithione) complexes within the density functional approach. *Theor. Chem. Acc.* **2007**, *117*, 621–635. [CrossRef]

26. Visentini, G.; Masino, M. Experimental determination of BEDT-TTF electron-molecular vibration constants through optical microreflectance. *Phys. Rev. B* **1998**, *58*, 9460–9467. [CrossRef]

27. Tajima, H.; Yakushi, K.; Kuroda, H. Polarized reflectance spectrum of β-(BEDT-TTF)$_2$I$_3$ single crystal. *Solid State Commun.* **1985**, *56*, 159–163. [CrossRef]

28. Yamamoto, T.; Uruichi, M.; Yamamoto, K.; Yakushi, K.; Kawamoto, A.; Taniguchi, H. Examination of the Charge-Sensitive Vibrational Modes in Bis(ethylenedithio)tetrathiafulvalene. *J. Phys. Chem. B* **2005**, *109*, 15226–15235. [CrossRef] [PubMed]

29. Yamamoto, T.; Nakazawa, Y.; Tamura, M.; Fukunaga, T.; Kato, R.; Yakushi, K. Vibrational Spectra of [Pd(dmit)$_2$] Dimer (dmit = 1,3-dithiole-2-thione-4,5-dithiolate): Methodology for examining charge, inter-molecular interactions, and orbital. *J. Phys. Soc. Jpn.* **2011**, *80*. [CrossRef]

30. Jacobsen, C.S.; Tanner, D.B. Electronic structure of some *p*-(C$_{10}$H$_8$S$_8$)2X compounds as studied by infrared spectroscopy. *Phys. Rev. B.* **1987**, *35*, 9605–9612. [CrossRef]

31. Świetlik, R.; Połomska, M.; Ouahab, L.; Guillevic, J. Infrared and Raman studies of the k-phase charge-transfer salts formed by BEDT-TTF and magnetic anions M(CN)$_6$$^{3-}$ (where M = Co^{III}, Fe^{III}, Cr^{III}). *J. Mater. Chem.* **2001**, *11*, 1313–1318. [CrossRef]

32. Sugimoto, T.; Fujiwara, H.; Noguchi, S.; Murata, K. New aspects of π–d interactions in magnetic molecular conductors. *Sci. Technol. Adv. Mater.* **2009**, *10*, 024302. [CrossRef] [PubMed]

33. Alemany, P.; Pouget, J.-P.; Canadell, E. Structural and electronic control of the metal to insulator transition and local orderings in the θ-(BEDT-TTF)2X organic conductors. *J. Phys. Condens. Matter* **2015**, *27*, 465702. [CrossRef] [PubMed]

34. Kahn, O. *Molecular Magnetism*; VCH Publishers: Weinheim, Germany, 1993.

35. Kinoshita, M.; Novoa, J.J.; Inoue, K.; Rawson, J.M.; Arčon, D. *π-Electron Magnetism. From Molecules to Magnetic Materials*; Veciana, J., Ed.; Springer: Berlin/Heidelberg, Germany, 2001.

36. Tian, H.; Wei, S.; Zheng, N.; Bo, N.; Peng, C. Magnetic blocking from exchange interactions: Slow relaxation of the magnetization and hysteresis loop observed in a dysprosium–nitronyl nitroxide chain compound with an antiferromagnetic ground state. *Chem. Eur. J.* **2013**, *19*, 994–1001.

37. Yatoo, M.A.; Cosquer, G.; Morimoto, M.; Irie, M.; Breedlove, B.K.; Yamashita, M. 1D chains of lanthanoid ions and a dithienylethene ligand showing slow relaxation of the magnetization. *Magnetochemistry* **2016**, *2*, 21. [CrossRef]

38. Zeng, D.; Ren, M.; Bao, S.-S.; Zheng, L.M. Tuning the coordination geometries and magnetic dynamics of [Ln(hfac)$_4$]$^-$ through alkali metal counterions. *Inorg. Chem.* **2014**, *53*, 795–801. [CrossRef] [PubMed]

39. *Crystal Clear-SM*, 1.4.0 SP1; Rigaku Corporation: Tokyo, Japan, 17 April 2008.

40. Altomare, A.; Burla, M.C.; Camalli, M.; Cascarano, G.L.; Giacovazzo, C.; Guagliardi, A.; Moliterni, A.G.G.; Polidori, G.; Spagna, R. SIR97: A new tool for crystal structure determination and refinement. *J. Appl. Crystallogr.* **1999**, *32*, 115–119. [CrossRef]

41. Farrugia, L.J. WinGX and ORTEP for Windows: An update. *J. Appl. Crystallogr.* **2012**, *45*, 849–854. [CrossRef]

42. Sheldrick, G.M. Crystal structure refinement with SHELXL. *Acta Cryst. C* **2015**, *71*, 3–8. [CrossRef] [PubMed]

magnetochemistry

MDPI

Article

Magneto-Luminescence Correlation in the Textbook Dysprosium(III) Nitrate Single-Ion Magnet

Ekaterina Mamontova [1], Jérôme Long [1,*], Rute A. S. Ferreira [2], Alexandre M. P. Botas [2,3], Dominique Luneau [4], Yannick Guari [1], Luis D. Carlos [2] and Joulia Larionova [1]

[1]   Institut Charles Gerhardt Montpellier, UMR 5253, Ingénierie Moléculaire et Nano-Objects, Université de Montpellier, ENSCM, CNRS, Place E. Bataillon, 34095 Montpellier CEDEX 5, France; ekaterina.mamontova@etu.umontpellier.fr (E.M.); yannick.guari@umontpellier.fr (Y.G.); joulia.larionova@umontpellier.fr (J.L.)

[2]   Physics Department and CICECO—Aveiro Institute of Materials, University of Aveiro, 3810-193 Aveiro, Portugal; rferreira@ua.pt (R.A.S.F.); a.botas@ua.pt (A.M.P.B.); lcarlos@ua.pt (L.D.C.)

[3]   I3N, University of Aveiro, 3810-193 Aveiro, Portugal

[4]   Laboratoire des Multimatériaux et Interfaces (UMR 5616), Université Claude Bernard Lyon 1, Campus de la Doua, 69622 CEDEX Villeurbanne, France; dominique.luneau@univ-lyon1.fr

*   Correspondence: jerome.long@umontpellier.fr; Tel.: +33-4-67-14-38-33

Academic Editor: Kevin Bernot

Received: 11 October 2016; Accepted: 12 November 2016; Published: 18 November 2016

Abstract: Multifunctional Single-Molecule Magnets (SMMs) or Single-Ion Magnets (SIMs) are intriguing molecule-based materials presenting an association of the slow magnetic relaxation with other physical properties. In this article, we present an example of a very simple molecule based on Dy^{3+} ion exhibiting a field induced SIM property and a characteristic Dy^{3+} based emission. The $[Dy(NO_3)_3(H_2O)_4] \cdot 2H_2O$ (1) complex is characterized by the means of single crystal X-Ray diffraction and their magnetic and photo-luminescent properties are investigated. We demonstrate here that it is possible to correlate the magnetic and luminescent properties and to obtain the Orbach barrier from the low temperature emission spectra, which is often difficult to properly extract from the magnetic measurements, especially in the case of field induced SIMs.

Keywords: lanthanides; single-ion magnets; single-molecule magnet; luminescence; magnetic relaxation

1. Introduction

Single-Ion Magnets (SIMs) are intriguing mononuclear coordination complexes which represent one of the smallest magnetically bistable units since they are composed of a single paramagnetic transition metal or lanthanide ion [1–3]. Consequently, such molecular objects appear as promising candidates for high-density data storage or quantum computing [1,4]. In the case of lanthanide ions, the magnitude of the anisotropic barrier responsible for the reversal of the magnetization arises from the intrinsic electronic structure and in particular from the sub-levels structure of the Stark components. Such energetic considerations rely on numerous parameters including the nature of the lanthanide ion (Kramers/non-Kramers character, oblate/prolate electronic density), its geometry as well as the overall coordination environment often governed by the donor strength of the ligands [5]. In addition, the relaxation of the magnetization may be governed by various mechanisms that are either spin-lattice processes (Orbach, Raman and direct relaxations) or the Quantum Tunneling of the Magnetization (QTM). The latter is known to depend also on numerous parameters such as the occurrence of magnetic exchange interactions, hyperfine coupling or dipolar interactions [6]. Understanding the mechanisms that govern the relaxation of the magnetization and the parameters that affect these relaxation processes is particularly important to ultimately optimize the SIM features.

In this line of scope, luminescent SIMs are particularly appropriate for these investigations since the photo-luminescence measurements permit determining spectroscopically the Stark sub-level structure of the lanthanides ions and comparing it with the magnetic data. However, examples of such magneto-optical correlations are rather scarce [7–18] in comparison with the large number of reported lanthanides-based SIMs. This fact is often due to the weak or completely quenched emission in these complexes. In order to overcome this problem, antenna ligands are used to enhance the lanthanide emission. Thus, in all cases of magneto-luminescent SIM, the observed lanthanide luminescence relies on the use of coordinated ligands or complexes as sensitizers to enhance the lanthanide luminescence via an energy transfer mechanism [19]. We report in this article that such magneto-optical correlations can be extended to a textbook dysprosium(III) nitrate coordination complex by direct excitation in the f–f transitions of the lanthanide ion. Surprisingly, this complex behaves as a field induced SIM allowing to perform a magneto-optical correlation.

2. Results

2.1. Structure

X-Ray Single Crystal diffraction indicates that $[Dy(NO_3)_3(H_2O)_4] \cdot 2H_2O$ (**1**) crystallizes in the triclinic *P*-1 space group (Table S1) with a unique crystallographic molecule. The coordination sphere of the Dy^{3+} ion is constituted by three bidentate nitrate and four water molecules leading to an overall coordination number of 10 (Figure 1). The Dy–O distances are comprised between 2.357(6) and 2.792(5) Å. The analysis of the geometry using the SHAPE software [20] indicates that it is close to a spherocorona (Table S2). Study of the crystal packing reveals a complex hydrogen bonds network involving oxygen nitrates and both coordinated and uncoordinated water molecules giving rise to a three-dimensional supramolecular network. The shortest Dy–Dy distance in the crystal is found to be 6.3282(7) Å.

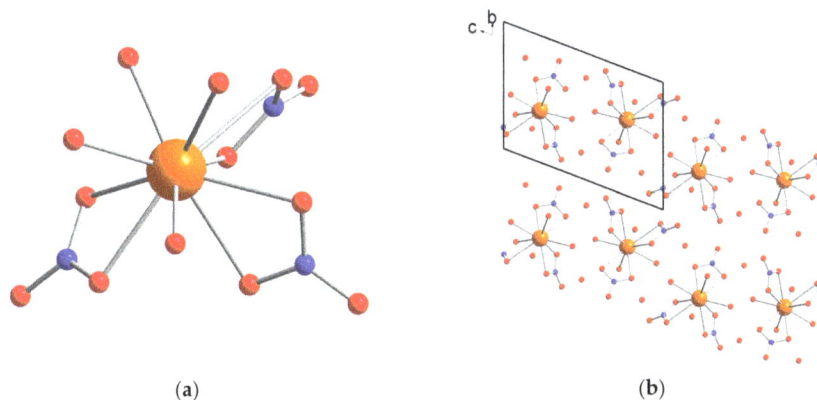

(a)   (b)

Figure 1. (**a**) molecular structure of **1**. Color code: **orange**, Dy; **red**, O; **blue**, N; **light grey**, C. Hydrogen atoms are omitted for clarity; and (**b**) view of the packing arrangement along the *a* crystallographic axis.

2.2. Magnetic Properties

The magnetic properties of **1** were investigated using a Superconducting QUantum Interference Device (SQUID) magnetometer (Quantum Design, San Diego, CA, USA) working in the temperature range 1.8–350 K up to 7 T in direct current (DC) and alternating currents (AC) modes.

2.2.1. DC Magnetic Properties

The room temperature value of χT is equal to 14.04 cm^3·K·mol^{-1}, which is in a good agreement with the value of 14.17 cm^3·K·mol^{-1} expected for a Dy^{3+} ion using the free-ion approximation. Upon cooling, χT remains constant down to 50 K before decreasing to reach the value of 11.87 cm^3·K·mol^{-1} at 1.8 K (Figure 2). Such a phenomenon can be ascribed to the thermal depopulation of the Stark sublevels. The field dependence of the magnetization measured at 1.8 K reaches the value of 5.85 μ_B under 70 kOe without evidence of a clear saturation (inset of Figure 2).

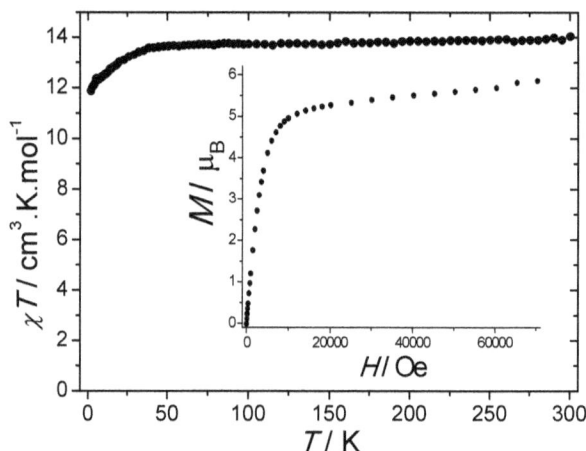

Figure 2. Temperature dependence of χT measured under a 1000 Oe DC field. Inset: field dependence of the magnetization measured at 1.8 K.

2.2.2. AC Magnetic Properties

AC magnetic properties were performed in order to investigate the occurrence of a slow relaxation of the magnetization. Under a zero DC field, no significant out-of-phase susceptibility, χ'', can be observed. This directly reflects a strong QTM which may arise from external factors such as dipolar interactions, intrinsic effects due to hyperfine interactions or symmetry deviations [6]. Applying DC fields induce a short-cut of this QTM and allow the observation of an out-of-phase component. The highest relaxation time, τ, is found at 4.0 K around 1500–2000 Oe DC fields (Figure 3). For higher field values, a decrease of τ is observed and can be imputed to the direct process that becomes predominant. Thermally activated and Raman processes are not field dependent but cannot be neglected even at 4.0 K, and the field dependence of the relaxation time can be fitted by using the following equation [21]:

$$\tau^{-1} = DH^4T + B_1/(1 + B_2H^2) + \tau_0^{-1}\exp(-\Delta/kT) + CT^m. \tag{1}$$

The first term accounts for the direct process (for Kramers-ion), the second one stands for the QTM while the third and fourth are relative to thermally activated and Raman processes, respectively. The best fit leads to the following values: $D = 1.37 \times 10^{-12}$ s^{-1}·K^{-1}·Oe^{-4}; $B_1 = 293$ s^{-1}, $B_2 = 2.6 \times 10^{-6}$ Oe^{-2}; $C = 0.0005$ s^{-1}·K^{-9}; $\Delta = 39$ cm^{-1}; $\tau_0 = 8 \times 10^{-10}$ s. The magnitudes of the B_1 and B_2 parameters directly reflect the degree of mixing between the $\pm m_J$ levels and consequently are a sign of the magnitude of the QTM process.

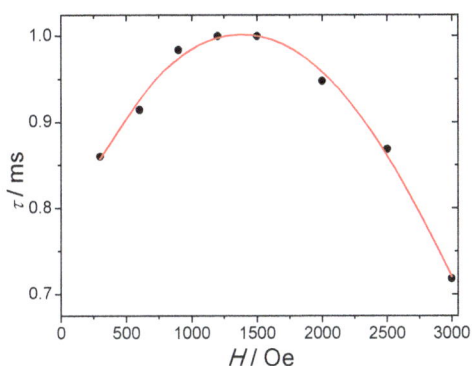

Figure 3. Field dependence of the relaxation time measured at 4.0 K. The **red** solid line represents the fit using Equation (1).

The frequency dependence of χ'' for various temperatures under a 2000 Oe DC field reveals a frequency dependent asymmetric peak (Figure 4a). The maximum of the peak shifts to higher temperatures upon increasing frequency, which indicates a field-induced slow relaxation of the magnetization. At low temperature, a plateau at low frequencies can be observed demonstrating the occurrence of a second relaxation process. We note that this process is essentially frequency independent, suggesting that it results from a collective behavior mediated by dipolar interactions and/or hydrogen bond network. As regards the frequency dependent process, it can be noticed that the magnitude of the out-of-phase signals increases with frequency, which is usually ascribed to the presence of intermolecular interactions or spin-glass behavior [22]. The same feature can also be observed on the temperature dependence of the out-of-phase susceptibility with various frequencies (Figure 4b).

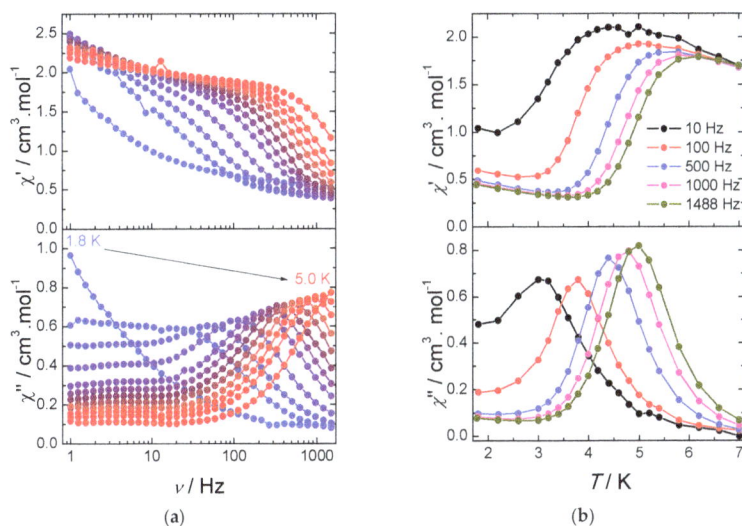

Figure 4. (**a**) frequency dependence of the in-phase (χ') and out-of-phase (χ'') components of the AC susceptibility under an optimal magnetic DC field of 2000 Oe for **1**; and (**b**) temperature dependence of the in-phase (χ') and out-of-phase (χ'') components of the AC susceptibility under a 2000 Oe DC field for **1**. Solid lines are guides to the eye.

The presence of a second relaxation process at a low temperature becomes evident when looking at the Cole–Cole plots (Figure 5a). Although a second semi-circle is not observed, a plateau is present. Therefore, the Cole–Cole plots were fitted by taking into account only the data of the semi-circle corresponding to the high-temperature process (Table S3). The values of the α parameter are ranging between 0.114 and 0.227, indicating a narrow distribution of the relaxation processes. Further insights into the mechanisms of the slow relaxation of the magnetization can be obtained by studying the temperature dependence of τ (Figure 5b). It turns out that a deviation from the thermally activated behavior can be observed at low temperatures and originates from the occurrence of other relaxation processes, such as Raman and/or direct. Fitting the temperature dependence of the relaxation time was performed using the following model: [23]

$$\tau^{-1} = \tau_0^{-1}\exp(-\Delta/kT) + CT^m + AT^n, \tag{2}$$

for which the first term accounts for a thermally activated process such as Orbach or direct spin–phonon transitions involving higher excited states [24], while the second and third ones stand for two-phonon Raman and direct relaxation processes, respectively. To avoid the over-parameterization, the values of $m = 9$ and $n = 1$ were fixed to the values found for two-phonon Raman (for Kramers ions) and direct processes [25,26].

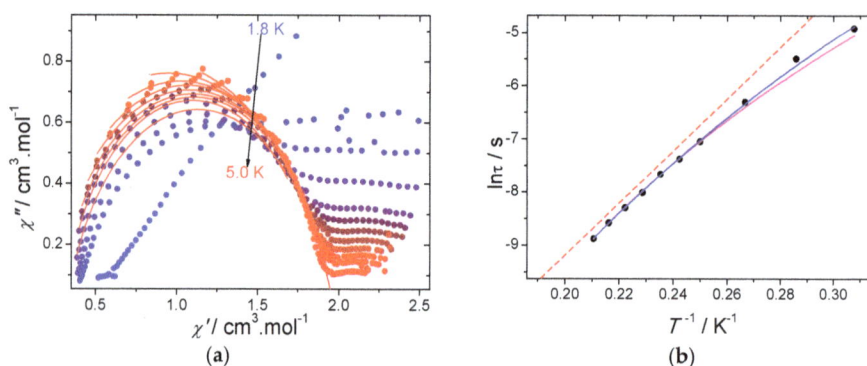

Figure 5. (**a**) Cole–Cole plots using the AC data performed under a 2000 Oe DC field. The **red** solid lines correspond to the fit with a generalized Debye model; (**b**) temperature dependence of the relaxation time using the AC data under a 2000 Oe DC field. The **blue** solid line corresponds to the fit with Equation (2), the **purple** solid line represents the fit with Equation (2) by fixing the Δ_{Orbach} parameter from photoluminescence, while the **red** dashed line represents the Orbach process and (Δ_{Orbach} fixed from photoluminescence and τ_0 obtained from Equation (2)).

The best fit parameters give $\Delta = 40 \pm 4$ cm^{-1}; $\tau_0 = (1.38 \pm 0.09) \times 10^{-9}$ s and $C = 0.003 \pm 0.001$ s$^{-1}\cdot$K^{-9} while the value for A is found negligible. Consequently, the deviation from linearity observed in Figure 5b can be explained by the overlap between thermally activated and Raman processes at low temperature. Fitting the temperature dependence of the relaxation time with only a Raman process (m being free) did not provide meaningful results with an m exponent value larger (10.7) than the expected value for the Kramers ion. Consequently, the thermally activated relaxation process seems to dominate in the high-temperature region.

2.3. Luminescence

The photoluminescence properties of **1** were investigated at 14 K and at 300 K, Figure 6a,b. Independently of the temperature, the emission spectra reveal the characteristic luminescence of the Dy^{3+} ions ascribed to the $^4F_{9/2}\rightarrow\,^6H_{15/2-11/2}$ transitions. By decreasing the temperature from 300 K to

14 K, a decrease in the full-width-at-half-maximum (fwhm) is observed, as well as the Stark splitting of the intra-4f^9 lines. The excitation spectra monitored within the more intense $^4F_{9/2}\rightarrow{}^6H_{13/2}$ transition reveal a series of intra-4f^9 straight lines attributed to transitions between the $^4K_{15/2,17/2}$, $^4I_{9/2-13/2}$, $^4G_{9/2-11/2}$, $^4M_{15/2-21/2}$, $^6P_{3/2-7/2}$, $^4F_{7/2-9/2}$, and $^4H_{15/2}$ excited levels and the $^6H_{11/2}$ ground multiplet. Similarly to what is found for the emission spectra, apart from a decrease of the fwhm, the intra-4f^9 lines, the excitation spectrum observed at 14 K resembles that measured at 300 K. The $^4F_{9/2}$ emission decay curves were measured under UV excitation at 355 nm (Figure S1) for **1**. Both decay curves are well-described by a single-exponential function yielding lifetime values of $\tau = (7.7 \pm 0.2) \times 10^{-9}$ s (14 K) and $\tau = (5.9 \pm 0.3) \times 10^{-9}$ s (300 K).

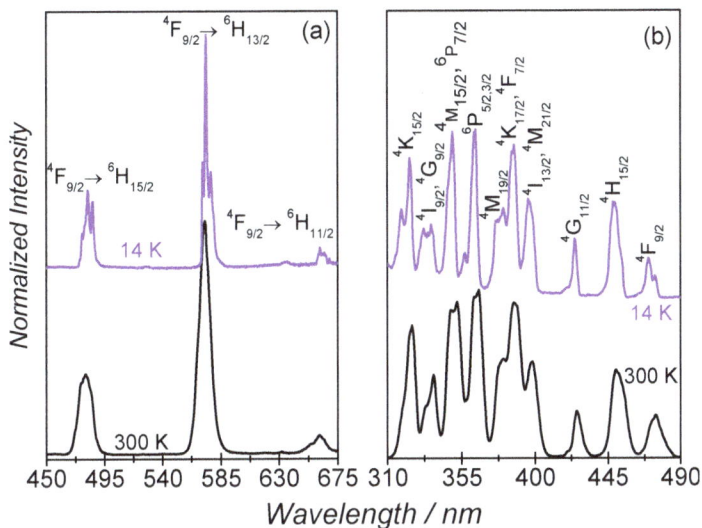

Figure 6. (**a**) emission; and (**b**) excitation spectra (14 K and 300 K) excited at 385 nm and monitored at 573 nm, respectively. In (**b**), the $^6H_{11/2}$ ground state is omitted and only the excited state is represented for clarity.

In order to gain additional insight into the correlation between SIM behaviour and luminescence properties, the crystal field splitting of the ground state of the Dy(III) ion in **1** was studied. The low temperature (14 K) high-resolution emission spectrum in the spectral region showing the $^4F_{9/2}\rightarrow{}^6H_{15/2}$ transitions was acquired for several crystals (Figure 7 and Figure S2). Apart from changes in the relative intensity of the Stark components, all the spectra reveal the presence of 11 components, as illustrated in Figure 7 for a selected ensemble of crystals.

Attending to the fact that the Dy(III) ions occupy a low-symmetry group, the splitting of the electronic levels ($^4F_{9/2}$ and $^6H_{15/2}$) into the maximum number of allowed components to $(2J + 1)/2$ Kramers doublets is expected, leading to five and eight components for $^4F_{9/2}$ and $^6H_{15/2}$, respectively. Thus, at least eight Stark components are expected for the $^4F_{9/2}\rightarrow{}^6H_{15/2}$ transitions assuming that only the lower-energy Stark component of the $^4F_{9/2}$ excited state is populated and all the transitions end at the Stark components of the ground multiplet $^6H_{15/2}$. Since 11 components are clearly expressed, it readily points out that the second Stark component of the $^4F_{9/2}$ excited state must also be populated. This is a feasible situation due to the low-energetic difference between the first Stark component and the second one (35 ± 5 cm^{-1}) which enables the population of the second Stark component of the $^4F_{9/2}$ at 14 K. The high-resolution emission spectrum was fitted with 11 components using Gaussian functions, whose energy was constrained to the peak position analysis based on the spectrum observation taking into account the experimental uncertainty (± 5 cm^{-1}); the fwhm and the relative intensity were free to vary. From the emission spectra best fit, an energy diagram of the Stark-sub-levels is proposed in Figure 7d. From the data acquired for the three distinct ensemble of crystals (Figure S2), an average energy barrier, $\Delta_{Orbach} = 34 \pm 5$ cm^{-1} is estimated, which is in excellent agreement with that obtained from the AC magnetic susceptibility data.

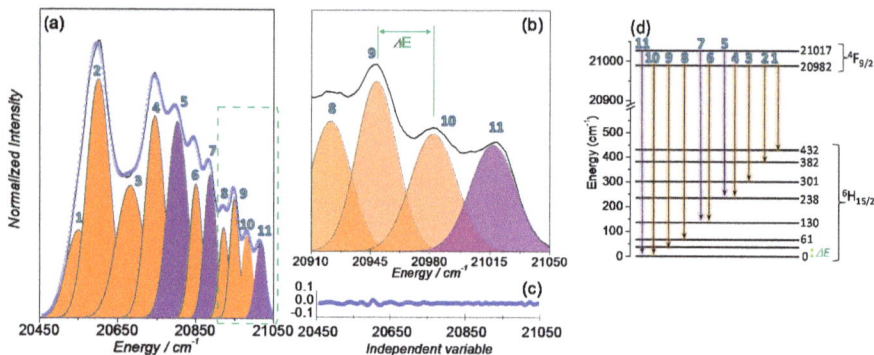

Figure 7. (**a,b**) magnification of the $^4F_{9/2} \rightarrow ^6H_{15/2}$ transition at 14 K and excited at 385 nm. Multi-Gaussian functions envelope fit (circles) and the components arising from the (**orange** shadow) first and (**purple** shadow) second $^4F_{9/2}$ Stark sublevels to the $^6H_{15/2}$ multiplet; (**c**) fit regular residual plot; and (**d**) schematic diagram of the radiative transitions between the Stark sublevels of the $^4F_{9/2}$ and $^6H_{15/2}$ multiplets of the Dy(III) ion.

3. Discussion

Compound **1** behaves as a field induced SIM showing two relaxation processes. Multiple relaxation processes have been widely observed in lanthanide zero-field or field induced SIMs [3,27,28]. For **1**, the independent frequency process observed in the low frequency range most likely arises from a collective behavior resulting from strong dipolar interactions or hydrogen bond network involving water molecules. Such intermolecular interactions may also affect the frequency dependent process associated with a slow relaxation of the magnetization. Indeed, a clear increase of the out-of-phase signal are observed with increasing frequencies. This fact is frequently viewed as the signature for either spin-glass or superparamagnetic systems with strong intermolecular interactions [29]. Dilution into a diamagnetic matrix would have been the ultimate proof to confirm the origin of the frequency independent relaxation process. However, attempts to grow crystals of a solid solution of Y^{3+}/Dy^{3+} ions were unsuccessful.

The field-induced SIM character of **1** appears at first glance rather surprising regarding the coordination environment of the dysprosium site. Based on the electrostatic proposed by Rinehardt and Long [5] and later extended by Chilton et al. [30], the stabilization of the oblate electronic density of the Dy^{3+} ion requires an axial crystal field. Although this model only relies on the charge of the atoms belonging to the ligands (it does not take into account the donor strength), it allows for estimating the orientation of the anisotropic axis by using the MAGELLAN software [30]. The orientation of the latter is highly dependent on the magnitude of the interaction between the negatively charged atom of the ligand and the Dy^{3+} ion. It turns out that the anisotropic axis (Figure 8) is found nearly collinear (deviation of 23.62° with respect to the central nitrogen atom) to the nitrate anion presenting the shortest Dy–O distance (Dy–O8 = 2.463 Å). The ideal stabilization of oblate density would have been to place two nitrate in *trans* fashion. However, the three nitrate are located in *cis* fashion to each others, forming a pseudo-triangular arrangement. We note also that the second shortest Dy–O4 = 2.491 Å distance forms an angle of 32.48° with the anisotropic axis. The deviation of the anisotropic axis with respect to the O4–Dy–O8 bond most likely relies on the effect of the positively charged nitrogen that attracts the electron density of the lanthanide ions. Significant longer Dy–O bonds (Dy–O1 2.529 Å and Dy–O5 = 2.568 Å) are found almost perpendicular to the anisotropic axis in order to limit the repulsion with the radial plane of the oblate electron density from the Dy^{3+} ion.

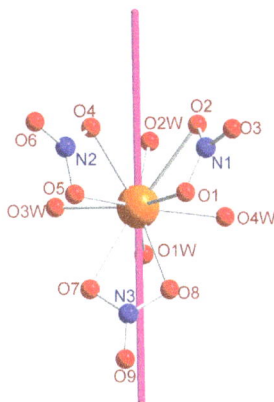

Figure 8. Arrangement of the anisotropic axis (**purple**) obtained with the MAGELLAN software.

Photoluminescence studies show that the characteristic emission of Dy^{3+} ion in **1** can be obtained without any sensitizer ligands by direct excitation into the f–f transitions. To our knowledge, this constitutes the first example of a luminescent SIM showing such an excitation process. The magneto-optical correlation shows that Δ obtained by magnetism (i.e., 40 ± 4 cm^{-1}) is very close to the energy gap between the ground and first-excited Kramers doublet extracted from photoluminescence measurements (i.e., 34 ± 5 cm^{-1}), indicating that the dominant relaxation pathways in the high temperature region is of the Orbach type. The observed slight deviation most likely arises from the overlap between Orbach and Raman processes due to the limited frequency of the oscillating field by standard SQUID magnetometry (i.e., 1488 Hz), even in the highest temperature range. The energy gap between the ground and first-excited Kramers doublet obtained from photoluminescence corresponds definitely to the correct value. Consequently, fitting of the temperature dependence of the relaxation time with Equation (2) can be performed by fixing Δ_{Orbach} to the one obtained by luminescence (~34 cm^{-1}). The best fit parameters obtained are $\tau_0 = (5.7 \pm 0.4) \times 10^{-9}$ s and $C = 0.0009 \pm 0.0003$ s$^{-1} \cdot$K^{-9}, while the value for A is still negligible. Attempts to let the m parameter free does not improve significantly the quality of the fit, and the m parameter is found slightly higher than its theoretical value of nine (fit $m = 10.7$) [31]. From these results, the C parameter is found lowered with respect to the fit without fixing Δ_{Orbach}. This confirms that the main source of relaxation is clearly an Orbach process (red dashed line, Figure 5b). This constitutes a rare example of a luminescent SIM without an underestimation of Δ obtained by magnetism as frequently observed in luminescent SIMs [7–18].

4. Materials and Methods

4.1. Synthesis and Crystal Structure

The compound was purchased from Alfa Aesar and used as received. Single crystals of 1 were selected and measured on an Xcalibur, Onyx diffractometer (Oxford Diffraction, Oxford, UK). The crystal was kept at 300 K during data collection. Using Olex2 [32], the structure was solved with the Superflip [33] structure solution program using Charge Flipping and refined with the olex2.refine [34] refinement package using Gauss–Newton minimisation.

4.2. Magnetic Measurements

Magnetic susceptibility data were collected with a Quantum Design (San Diego, CA, USA) MPMS-XL SQUID magnetometer working between 1.8 and 350 K with the magnetic field up to 7 Tesla. The data were corrected for the sample holder and the diamagnetic contributions calculated from Pascal's constants. The AC magnetic susceptibility measurements were carried out in the presence of a 3 Oe oscillating field in zero or applied external DC field.

4.3. Photoluminescence Measurements

The photoluminescence spectra were recorded at 14 K and at 300 K with a modular double grating excitation spectrofluorimeter with a TRIAX 320 emission monochromator (Fluorolog-3, Horiba Scientific (Kyoto, Japan)) coupled to a R928 photomultiplier (Hamamatsu) using a front face acquisition mode. The excitation source was a 450 W Xe arc lamp. All the emission spectra were measured under vaccum conditions (10^{-7} bar) and corrected for detection and optical spectral response of the spectrofluorimeter, and the excitation spectra were corrected for the spectral distribution of the lamp intensity using a photodiode reference detector. The nominal spectral dispersion response of the spectrofluorimeter is 2.64 nm·cm^{-1}, yielding a spectral resolution of 0.1 nm (± 5 cm^{-1}) for the high-resolution emission spectra (slits width of 5×10^{-2} mm). The emission decay curves were measured at 14 K and at 300 K using a pulsed LED peaking at 355 nm (SpectraLED-355, Horiba Scientific (Kyoto, Japan)) coupled to a TBX-04 photomultiplier tube module. The emission spectrum of the SpectraLED-355 displays a Gaussian profile with a fwmh of ca. 15 nm. The room temperature emission quantum yield was measured using the C9920-02 (Hamamatsu) setup with a 150 W Xe lamp coupled to a monochromator for wavelength discrimination, an integration sphere as sample chamber and a multichannel analyzer for signal detection. The accuracy is within 10% according to the manufacturer. Three measurements were made for each sample, revealing values below the detection limits of our experimental apparatus (0.01).

5. Conclusions

Although known for decades and thoroughly used for the synthesis of SIMs and SMMs, we have demonstrated that the dysprosium(III) nitrate salt, $[Dy(NO_3)_3(H_2O)_4]\cdot 2H_2O$ (**1**), behaves as a field-induced SIM. This can be rationalized by the presence of the negatively charged oxygen of the nitrate ions, which helps to stabilize the oblate electronic density of the Dy^{3+} ion. In addition, the excitation in the intra 4f transitions induces the characteristic luminescence of the Dy^{3+} ion and allows to perform a magneto-structural correlation, giving additional insights into the relaxation processes. In the high-temperature region, it proceeds via an Orbach mechanism, which overlaps with the Raman process upon lowering of the temperature.

Supplementary Materials: The following are available online at www.mdpi.com/2312-7481/2/4/41/s1, Table S1: Crystal and structure refinement data for compound 1; Table S2: SHAPE analysis for compound 1; Table S3: Fitting of the Cole–Cole plots with a generalized Debye model for temperature ranging from 1.8 to 4.1 K under a 2000 Oe DC field for 1; Figure S1: Emission decay curves; Figure S2: High-resolution emission spectra in the $^4F_{9/2} \rightarrow ^6H_{15/2}$ transition acquired for distinct ensemble of crystals.

Acknowledgments: The authors thank the University of Montpellier, Centre Nationale de la Recherche Scientifique (CNRS) and Plateforme d'Analyse et de Caractérisation (PAC) Balard Institut Charles Gerhardt Montpellier (ICGM). E.M. acknowledges the financial support from the LABoratoire d'EXcellence (LABEX) CheMISyst Agence Nationale de la Recherche ANR-10-LABX-05-01. This work was developed within the scope of the project CICECO-Aveiro Institute of Materials, POCI-01-0145-FEDER-007679 (FCT Ref. UID /CTM /50011/2013) and I3N (UID/CTM/50025/2013) financed by national funds through the FCT/MEC and co-financed by Fonds Européen de Développement Régional (FEDER) under the PT2020 Partnership. A.M.P.B. thanks FCT for a PhD fellowship (SFRH/BD/104789/2014). The Portugal–France bilateral action, PESSOA Program (Hubert Curien) "Multifunctional magneto-luminescent molecular architectures" is also acknowledged.

Author Contributions: All authors contributed equally to this work. Rute A. S. Ferreira, Alexandre M. P. Botas and Luis D. Carlos performed, interpreted and wrote the luminescence part; Dominique Luneau resolved the crystal structure; Ekaterina Mamontova and Jérôme Long performed the magnetic characterization; Jérôme Long wrote the draft of the article; Yannick Guari and Joulia Larionova discussed the results and implications and commented on the manuscript at all stages.

Conflicts of Interest: The authors declare no conflict of interest.

References

1. Woodruff, D.N.; Winpenny, R.E.P.; Layfield, R.A. Lanthanide single-molecule magnets. *Chem. Rev.* **2013**, *113*, 5110–5148. [CrossRef] [PubMed]
2. Layfield, R.A. Organometallic single-molecule magnets. *Organometallics* **2014**, *33*, 1084–1099. [CrossRef]
3. Liddle, S.T.; van Slageren, J. Improving f-element single molecule magnets. *Chem. Soc. Rev.* **2015**, *44*, 6655–6669. [CrossRef] [PubMed]
4. Luzon, J.; Sessoli, R. Lanthanides in molecular magnetism: So fascinating, so challenging. *Dalton Trans.* **2012**, *41*, 13556–13567. [CrossRef] [PubMed]
5. Rinehart, J.D.; Long, J.R. Exploiting single-ion anisotropy in the design of f-element single-molecule magnets. *Chem. Sci.* **2011**, *2*, 2078–2085. [CrossRef]
6. Pointillart, F.; Bernot, K.; Golhen, S.; Le Guennic, B.; Guizouarn, T.; Ouahab, L.; Cador, O. Magnetic memory in an isotopically enriched and magnetically isolated mononuclear dysprosium complex. *Angew. Chem. Int. Ed.* **2015**, *54*, 1504–1507. [CrossRef] [PubMed]
7. Cucinotta, G.; Perfetti, M.; Luzon, J.; Etienne, M.; Car, P.-E.; Caneschi, A.; Calvez, G.; Bernot, K.; Sessoli, R. Magnetic anisotropy in a dysprosium/dota single-molecule magnet: Beyond simple magneto-structural correlations. *Angew. Chem. Int. Ed.* **2012**, *51*, 1606–1610. [CrossRef] [PubMed]
8. Long, J.; Vallat, R.; Ferreira, R.A.S.; Carlos, L.D.; Almeida Paz, F.A.; Guari, Y.; Larionova, J. A bifunctional luminescent single-ion magnet: Towards correlation between luminescence studies and magnetic slow relaxation processes. *Chem. Commun.* **2012**, *48*, 9974–9976. [CrossRef] [PubMed]
9. Yamashita, K.; Miyazaki, R.; Kataoka, Y.; Nakanishi, T.; Hasegawa, Y.; Nakano, M.; Yamamura, T.; Kajiwara, T. A luminescent single-molecule magnet: Observation of magnetic anisotropy using emission as a probe. *Dalton Trans.* **2013**, *42*, 1987–1990. [CrossRef] [PubMed]
10. Pointillart, F.; Le Guennic, B.; Golhen, S.; Cador, O.; Maury, O.; Ouahab, L. A redox-active luminescent ytterbium-based single molecule magnet. *Chem. Commun.* **2013**, *49*, 615–617. [CrossRef] [PubMed]
11. Gavey, E.L.; Al Hareri, M.; Regier, J.; Carlos, L.D.; Ferreira, R.A.S.; Razavi, F.S.; Rawson, J.M.; Pilkington, M. Placing a crown on dy(III)—A dual property lniii crown ether complex displaying optical properties and smm behaviour. *J. Mater. Chem. C* **2015**, *3*, 7738–7747. [CrossRef]
12. Shintoyo, S.; Murakami, K.; Fujinami, T.; Matsumoto, N.; Mochida, N.; Ishida, T.; Sunatsuki, Y.; Watanabe, M.; Tsuchimoto, M.; Mrozinski, J.; et al. Crystal field splitting of the ground state of terbium(iii) and dysprosium(III) complexes with a triimidazolyl tripod ligand and an acetate determined by magnetic analysis and luminescence. *Inorg. Chem.* **2014**, *53*, 10359–10369. [CrossRef] [PubMed]
13. Ren, M.; Bao, S.-S.; Ferreira, R.A.S.; Zheng, L.-M.; Carlos, L.D. A layered erbium phosphonate in pseudo-D_{5h} symmetry exhibiting field-tunable magnetic relaxation and optical correlation. *Chem. Commun.* **2014**, *50*, 7621–7624. [CrossRef] [PubMed]
14. Long, J.; Rouquette, J.; Thibaud, J.-M.; Ferreira, R.A.S.; Carlos, L.D.; Donnadieu, B.; Vieru, V.; Chibotaru, L.F.; Konczewicz, L.; Haines, J.; et al. A high-temperature molecular ferroelectric Zn/Dy complex exhibiting single-ion-magnet behavior and lanthanide luminescence. *Angew. Chem. Int. Ed.* **2015**, *54*, 2236–2240. [CrossRef] [PubMed]
15. Rechkemmer, Y.; Fischer, J.E.; Marx, R.; Dörfel, M.; Neugebauer, P.; Horvath, S.; Gysler, M.; Brock-Nannestad, T.; Frey, W.; Reid, M.F.; et al. Comprehensive spectroscopic determination of the crystal field splitting in an erbium single-ion magnet. *J. Am. Chem. Soc.* **2015**, *137*, 13114–13120. [CrossRef] [PubMed]
16. Gregson, M.; Chilton, N.F.; Ariciu, A.-M.; Tuna, F.; Crowe, I.F.; Lewis, W.; Blake, A.J.; Collison, D.; McInnes, E.J.L.; Winpenny, R.E.P.; et al. A monometallic lanthanide bis(methanediide) single molecule magnet with a large energy barrier and complex spin relaxation behaviour. *Chem. Sci.* **2016**, *7*, 155–165. [CrossRef]

17. Bi, Y.; Chen, C.; Zhao, Y.-F.; Zhang, Y.-Q.; Jiang, S.-D.; Wang, B.-W.; Han, J.-B.; Sun, J.-L.; Bian, Z.-Q.; Wang, Z.-M.; et al. Thermostability and photoluminescence of dy(III) single-molecule magnets under a magnetic field. *Chem. Sci.* **2016**, *7*, 5026–5031. [CrossRef]

18. Al Hareri, M.; Gavey, E.L.; Regier, J.; Ras Ali, Z.; Carlos, L.D.; Ferreira, R.A.S.; Pilkington, M. Encapsulation of a $[Dy(OH_2)_8]^{3+}$ cation: Magneto-optical and theoretical studies of a caged, emissive smm. *Chem. Commun.* **2016**, *52*, 11335–11338. [CrossRef] [PubMed]

19. Bunzli, J.-C.G.; Piguet, C. Taking advantage of luminescent lanthanide ions. *Chem. Soc. Rev.* **2005**, *34*, 1048–1077. [CrossRef] [PubMed]

20. Casanova, D.; Llunell, M.; Alemany, P.; Alvarez, S. The rich stereochemistry of eight-vertex polyhedra: A continuous shape measures study. *Chem. Eur. J.* **2005**, *11*, 1479–1494. [CrossRef] [PubMed]

21. Zadrozny, J.M.; Atanasov, M.; Bryan, A.M.; Lin, C.-Y.; Rekken, B.D.; Power, P.P.; Neese, F.; Long, J.R. Slow magnetization dynamics in a series of two-coordinate iron(II) complexes. *Chem. Sci.* **2013**, *4*, 125–138. [CrossRef]

22. Bilyachenko, A.N.; Yalymov, A.I.; Korlyukov, A.A.; Long, J.; Larionova, J.; Guari, Y.; Zubavichus, Y.V.; Trigub, A.L.; Shubina, E.S.; Eremenko, I.L.; et al. Heterometallic Na_6Co_3 phenylsilsesquioxane exhibiting slow dynamic behavior in its magnetization. *Chem. Eur. J.* **2015**, *21*, 18563–18565. [CrossRef] [PubMed]

23. Meihaus, K.R.; Minasian, S.G.; Lukens, W.W.; Kozimor, S.A.; Shuh, D.K.; Tyliszczak, T.; Long, J.R. Influence of pyrazolate vs. *N*-heterocyclic carbene ligands on the slow magnetic relaxation of homoleptic trischelate lanthanide(III) and uranium(III) complexes. *J. Am. Chem. Soc.* **2014**, *136*, 6056–6068. [CrossRef] [PubMed]

24. Ungur, L.; Chibotaru, L.F. Strategies toward high-temperature lanthanide-based single-molecule magnets. *Inorg. Chem.* **2016**, *55*, 10043–10056. [CrossRef] [PubMed]

25. Shrivastava, K.N. Theory of spin–lattice relaxation. *Phys. Status Solidi B* **1983**, *117*, 437–458. [CrossRef]

26. Scott, P.L.; Jeffries, C.D. Spin-lattice relaxation in some rare-earth salts at helium temperatures; observation of the phonon bottleneck. *Phys. Rev.* **1962**, *127*, 32–51. [CrossRef]

27. Blagg, R.J.; Ungur, L.; Tuna, F.; Speak, J.; Comar, P.; Collison, D.; Wernsdorfer, W.; McInnes, E.J.L.; Chibotaru, L.F.; Winpenny, R.E.P. Magnetic relaxation pathways in lanthanide single-molecule magnets. *Nat. Chem.* **2013**, *5*, 673–678. [CrossRef] [PubMed]

28. Amjad, A.; Figuerola, A.; Caneschi, A.; Sorace, L. Multiple magnetization reversal channels observed in a 3d-4f single molecule magnet. *Magnetochemistry* **2016**, *2*. [CrossRef]

29. Mydosh, J.A. *Spin Glasses: An Experimental Introduction*; Taylor & Francis: London, UK; Washington, DC, USA, 1993.

30. Chilton, N.F.; Collison, D.; McInnes, E.J.L.; Winpenny, R.E.P.; Soncini, A. An electrostatic model for the determination of magnetic anisotropy in dysprosium complexes. *Nat. Commun.* **2013**, *4*. [CrossRef] [PubMed]

31. Orbach, R.; Blume, M. Spin-lattice relaxation in multilevel spin systems. *Phys. Rev. Lett.* **1962**, *8*, 478–480. [CrossRef]

32. Dolomanov, O.V.; Bourhis, L.J.; Gildea, R.J.; Howard, J.A.K.; Puschmann, H. Olex2: A complete structure solution, refinement and analysis program. *J. Appl. Cryst.* **2009**, *42*, 339–341. [CrossRef]

33. Palatinus, L.; Prathapa, S.J.; van Smaalen, S. Edma: A computer program for topological analysis of discrete electron densities. *J. Appl. Cryst.* **2012**, *45*, 575–580. [CrossRef]

34. Bourhis, L.J.; Dolomanov, O.V.; Gildea, R.J.; Howard, J.A.; Puschmann, H. The anatomy of a comprehensive constrained, restrained refinement program for the modern computing environment—Olex2 dissected. *Acta Cryst. A* **2015**, *71*, 59–75. [CrossRef] [PubMed]

magnetochemistry

MDPI

Article

Slow Magnetic Relaxation in Chiral Helicene-Based Coordination Complex of Dysprosium

Guglielmo Fernandez-Garcia [1], Jessica Flores Gonzalez [1], Jiang-Kun Ou-Yang [1], Nidal Saleh [1], Fabrice Pointillart [1], Olivier Cador [1], Thierry Guizouarn [1], Federico Totti [2], Lahcène Ouahab [1], Jeanne Crassous [1] and Boris Le Guennic [1,*]

[1] Institut des Sciences Chimiques de Rennes, UMR 6226 CNRS, Université de Rennes 1, 263 Avenue du Général Leclerc, 35042 Rennes CEDEX, France; guglielmo.fernandezgarcia@univ-rennes1.fr (G.F.-G.); jessica.flores-gonzales@univ-rennes1.fr (J.F.G.); jian.ou-yang@univ-rennes1.fr (J.-K.O.-Y.); nidal.saleh@cnrs.fr (N.S.); fabrice.pointillart@univ-rennes1.fr (F.P.); olivier.cador@univ-rennes1.fr (O.C.); thierry.guizouarn@univ-rennes1.fr (T.G.); lahcene.ouahab@univ-rennes1.fr (L.O.); jeanne.crassous@univ-rennes1.fr (J.C.)

[2] Department of Chemistry "Ugo Schiff" and INSTM RU, University of Florence, 50019 Sesto Fiorentino, Italy; federico.totti@unifi.it

[*] Correspondence: boris.leguennic@univ-rennes1.fr; Tel.: +33-02-23-23-35-21

Academic Editor: Kevin Bernot
Received: 9 November 2016; Accepted: 13 December 2016; Published: 23 December 2016

Abstract: The complex [Dy(**L**)(tta)$_3$] with **L** the chiral 3-(2-pyridyl)-4-aza[6]-helicene ligand (tta$^-$ = 2-thenoyltrifluoroaacetonate) has been synthesized in its racemic form and structurally and magnetically characterized. [Dy(**L**)(tta)$_3$] behaves as a single molecule magnet in its crystalline phase with the opening of a hysteresis loop at 0.50 K. These magnetic properties were interpreted with ab initio calculations.

Keywords: single molecule magnets; lanthanide; helicene; magnetic anisotropy; ab initio calculations

1. Introduction

The design of single molecule magnet (SMM), with the aim of enhancing its peculiar magnetic properties, has been a prolific field in the scientific community for decades [1–3]. Indeed, SMMs can pave the way towards a new generation of materials as, for example, molecular qubits for quantum computing [4], memory storage devices [5] or spin valves [6]. In this framework, lanthanide ions are commonly exploited in the effort of reaching slower relaxation rates for the reversal of the magnetization. Indeed, lanthanides are well classified by looking at their electron density distribution, ranging from oblate (planar) to prolate (axial) distribution [7]. This is mainly due to their strong spin-orbit coupling, which leads to ground states with large angular momentum J and strong magnetic anisotropy. The crystal field, induced by the donor atoms of the ligands, acts only as a perturbation on the electron density distribution, leading to a fine-tuning of the electronic properties and so of the molecular magnetism. As a consequence, the careful choice of the lanthanide ion and of the ligands (and the induced symmetry) can be used to engineer novel SMMs. However, a complete elucidation of these magneto-structural correlations for these complexes is still missing, even if progresses have been done recently [8–11].

For all the applications mentioned above, it is crucial to study the correlation between the SMM behavior and other physical properties such as luminescence or redox activity [12]. The versatility of ligand chemistry can be exploited in this sense and may offer the possibility to have in a single compound, for instance, a magnetic emitting nanodevice [13–15]. Indeed, lanthanides have been intensively studied for their peculiar luminescence that covers a broad range of frequencies (from visible to near IR) with sharp line shape emission bands and long lifetime of the excited states [16–22]. However, they show very low absorption coefficients, since the f–f transitions are indeed prohibited (Laporte rule) [23]. This results in ineffective direct excitation processes, especially in dilute solution. To tackle this problem, indirect sensitization, using for example MLCT (Metal-Ligand Charge Transfer) transitions, has been developed by means of ligands functionalization with organic chromophores acting as antennae [24]. In the case where the antenna is chiral, the solid-state properties might change between the enantiopure and the racemic crystals. Besides, the magnetic properties of these atoms can be used to modify the light absorption in chiral compounds, an effect known as magneto-chiral dichroism [25,26].

A first example of the coupling between a Dy^{III}-based SMM and a chiral antenna has been reported recently with the complex [Dy(**L**)(hfac)$_3$] with **L** = 3-(2-pyridyl)-4-aza[6]-helicene and hfac = 1,1,1,5,5,5-hexafluoroacetylacetone [27]. The Dy^{III} ion, with its $^6H_{15/2}$ ground state, easily leads to Ising type of magnetic anisotropy in coordination spheres like N_2O_6 and this is achieved with the common bidentate 2,2'-bipyridine (bpy) ligand and three hfac$^-$ ligands [13,28–30]. On the other hand, a 2,2'-bipyridine (bpy) ligand has been functionalized with a [6]-helicene to enhance the luminescence. Indeed, the latter presents a π-conjugated backbone of aromatic rings, configurationally stable for $n \geq 5$, and its peculiar topology results in intense emission [31,32]. Moreover, [n]-helicene ligands are helically-shaped, so they possess a chirality despite the absence of enantiocenter. Due to these intrinsic properties, this family of ligands is widely employed for various applications, ranging from organic molecular electronics [33], probes for detection of chirality and sensing devices [34] to molecular junction [35].

In the case of the [Dy(**L**)(hfac)$_3$] SMM [27], the chirality of the ligand results in two possible crystal structures (racemic and enantiopure) with similar molecular arrangement but different packings. Interestingly, racemic and enantiopure crystals show notable different magnetic behavior, with the opening of a magnetic hysteresis only in the case of the enantiopure. Moreover, the calculated different nature (antiferromagnetic and ferromagnetic) of the dipolar couplings between first-neighbors allows explaining the magnetic measurements (e.g., temperature dependence of $\chi_M T$).

With the aim to enhance the magnetic properties in this series of compounds, we present herein a novel derivative in which the hfac$^-$ ligands have been replaced by tta$^-$ (2-thenoyltrifluoroacetonate) ligands. Indeed, it is well known that the swapping of these two ligands in such specific N_2O_6 environment enhances the magnetic properties [36–38], even if in other coordination environments the opposite effect has been recently observed [39]. Therefore, we report the synthesis, the single crystal X-ray structural analysis and the magnetic characterization along with extensive ab initio calculations of the novel compound [Dy(**L**)(tta)$_3$] (**1**).

2. Results and Discussion

2.1. Structure

Complex **1** was obtained by the coordination reaction of the chiral 3-(2-pyridyl)-4-aza[6]-helicene [39,40] ligand (**L**) and tris(2-thenoyltrifluoroacetonate)bis(aqueous)Dy^{III} in CH_2Cl_2 (Scheme 1).

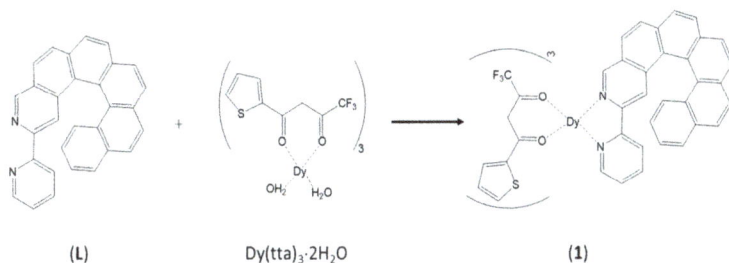

Scheme 1. Synthetic route to obtain complex **1**.

1 crystallizes in the triclinic centrosymmetric space group *P*-1 (Figure 1 and Figure S1, Table S1). The Dy^{III} ion is surrounded by two nitrogen atoms and six oxygen atoms coming from the three 2-thenoyltrifluoroacetonate (tta$^-$) anions and the **L** ligand. The N$_2$O$_6$ coordination polyhedron can be described as a distorted square antiprism environment (*D*$_{4d}$ symmetry on the basis of SHAPE analysis, Table S2) [41]. Thus, the replacement of the 1,1,1,5,5,5-hexafluoroacetylacetonate anions with tta$^-$ ones confers an higher symmetry for the coordination environment [27]. As already noted, such observation was already done for another complex stemming from this group and a significant positive impact on the magnetic properties was observed [37,38].

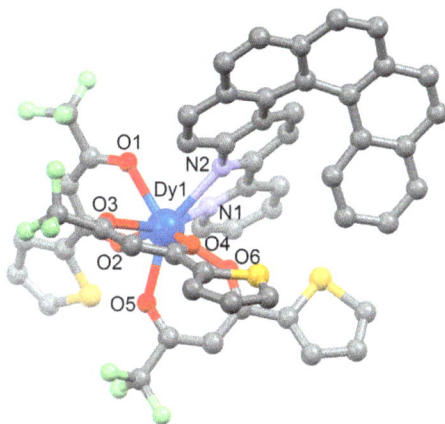

Figure 1. Molecular structure of **1**. Hydrogen atoms and molecules of crystallization are omitted for clarity. Selected bond lengths: Dy1–N1, 2.560(3) Å; Dy1–N2, 2.549(3) Å; Dy1–O1, 2.341(3) Å; Dy1–O2, 2.296(3) Å; Dy1–O3, 2.359(3) Å; Dy1–O4, 2.356(3) Å; Dy1–O5, 2.322(3) Å; Dy1–O6, 2.341(3) Å.

Starting from the racemic mixture of **L**, both enantiomers are present in the cell (*P*-1 space group symmetry). The crystal packing reveals that heterochiral dimers are formed with the presence of π–π interactions between the 2,2′-bipyridyl moieties (Figure 2) while an organic network runs along the *a*-axis thank to π–π interactions between the helicenic parts. The Dy–Dy shortest intermolecular distance was measured equal to 8.935 Å which is similar to the distance measured in the complex involving the Dy(hfac)$_3$ metallo-precursor.

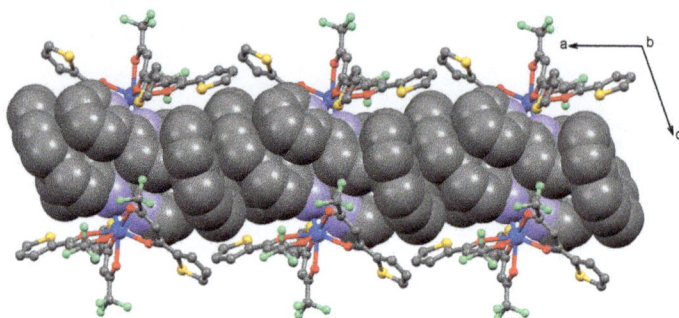

Figure 2. Crystal packing of **1** along the *a*-axis. "Spacefill" and "ball and sticks" representations are used for **L** ligands and organometallic moieties, respectively.

2.2. Magnetic Properties

2.2.1. Static Magnetic Measurements

The temperature dependence of $\chi_M T$ for the sample **1** is represented in Figure 3. The room temperature value is 13.96 cm^3·K·mol^{-1} in good agreement with the expected value of 14.17 cm^3·K·mol^{-1} for an isolated Dy^{III} ion [42]. Upon cooling, $\chi_M T$ decreases monotonically down to 11.20 cm^3·K·mol^{-1} due to the thermal depopulation of the M_J states. Below 5 K, the more rapid decrease could be attributed to the presence of weak dipolar antiferromagnetic interactions as determined by quantum calculations on the analogue complex involving hfac$^-$ ancillary ligands [27]. The field dependence of the magnetization measured at 2.0 K reaches the value of 5.12 Nβ under a magnetic field of 50 kOe, in agreement with the expected value (5 Nβ) for an Ising ground state (Inset of Figure 3).

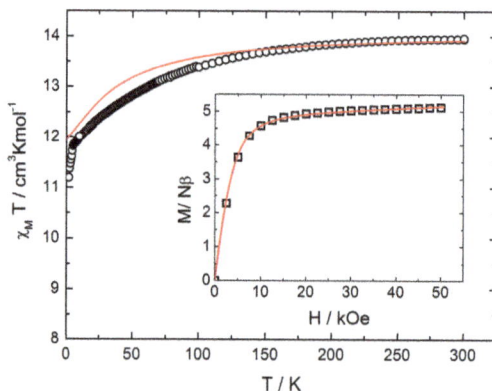

Figure 3. Temperature dependence of $\chi_M T$ for **1** (black circles). The inset shows the field variations of the magnetization at 2 K. Full red lines correspond to the simulated curves from ab initio calculations.

2.2.2. Dynamic Magnetic Measurements

The out-of-phase component of the ac susceptibility (χ_M'') of **1** was measured using immobilized crunched single crystals. It shows frequency dependence in zero external dc field with clear maxima on the χ_M'' vs. ν curves (ν the frequency of the ac oscillating field) up to 13 K (Figure 4a). The frequency dependence of the ac susceptibility can be analyzed in the framework of the extended Debye model

(Figures S2 and S3) [43,44]. The temperature dependence of the relaxation time at zero field is extracted between 2.0 and 14.0 K (Table S3). Formally, four different relaxation mechanisms coexist: Direct, Raman, Orbach and QTM [2]. The former disappears in the absence of external field while the second and the third are field-independent. The latest is the only temperature independent mechanism. Fitting of the zero-field data with only Raman and QTM is not satisfactory while Raman + Orbach + QTM leads to unrealistic results owing to over-parameterization. The only realistic picture is given by the Orbach + QTM combination. The relaxation time follows the Arrhenius law $\tau = \tau_0 \exp(\Delta/kT)$ only above 12 K with $\tau_0 = 2.6(2) \times 10^{-6}$ s and $\Delta = 38.7(2)$ cm^{-1} (Figure 4d, open squares) with dominant QTM mechanism ($\tau_{QTM} = 7.0(3) \times 10^{-4}$ s) at low temperature.

In order to reduce the QTM operating in this system, the optimal magnetic field of 1000 Oe was determined by a scan field (Figure 4b). The application of this moderate external dc field induces a slowing down of the magnetic relaxation with a shift of the maxima of the χ_M'' vs. ν curve at lower frequencies (Figure 4c). It must be mentioned that at moderate fields at least two relaxation processes coexist which merge into one at higher fields than 800 Oe (Figure 4b). Any attempts to fit the thermal behavior of the relaxation time with a combination of the previously mentioned mechanisms fail with a reasonable set of data. The thermal dependence of the relaxation time of the magnetization (Figures S4 and S5 and Table S4) can be fitted considering a combination of two thermally dependent regimes between 3.0 and 14.0 K (Orbach processes) ($\tau_0 = 2.0(7) \times 10^{-7}$ s and $\Delta_0 = 68.1(3)$ cm^{-1}, $\tau_1 = 4.9(6) \times 10^{-4}$ s and $\Delta_1 = 16.0(5)$ cm^{-1}). Relaxation times on the order of few seconds is slow enough to observe the opening of the hysteresis loop at 0.50 K (Figure 5) which remains opened at higher temperatures (Figure S6). One must mention that the hysteresis loop of the racemic form of [Dy(**L**)(hfac)$_3$] [27] was closed at the same temperature.

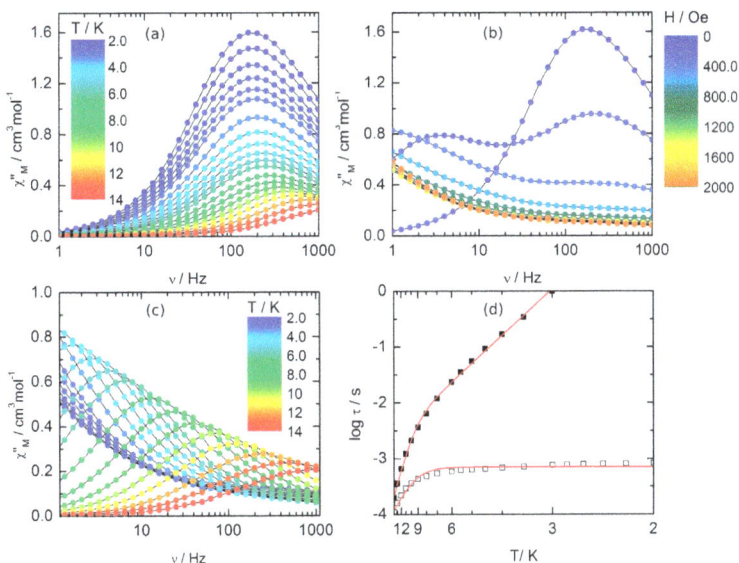

Figure 4. (**a**) Frequency dependence of χ_M'' between 2 and 14 K; (**b**) scan field of the frequency dependence of χ_M'' at 2 K; (**c**) frequency dependence of χ_M'' between 2 and 14 K under an applied magnetic field of 1000 Oe; and (**d**) temperature variation of the relaxation time measured in zero field (open squares) and in an external field of 1000 Oe (full squares) with the best fitted curve (red lines) in the temperature range of 2–14 K.

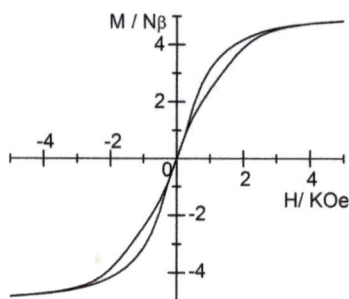

Figure 5. Magnetic hysteresis loop of **1** measured at 0.5 K at a sweep rate of 16 Oe·s^{-1}.

2.2.3. Ab Initio Calculations

Various theoretical models, with various pros and cons, are available to predict the magnetic properties of lanthanides, ranging from complete active space ab initio methods (e.g., CASSCF or CASPT2) to semi-empirical methods (e.g., radial effective charge model, REC) [45]. To study the electronic structure of the present compound, we choose SA-CASSCF/RASSI-SO calculations as a good compromise between accuracy with respect to the experimental evidence and "first principle" theoretical model.

Calculations were performed for **1** to understand the observed magnetic properties comparing with the ab initio calculated electronic structure (see computational details). The calculated $\chi_M T$ vs. T and M vs. H (Figure 3) curves fairly well reproduce the experimental curves, even if the agreement for the $\chi_M T$ vs. T data is still semi-quantitative. Calculations confirm the axial character of the magnetic anisotropy tensor of the ground Kramers doublet with large g_z values of 19.55 and almost negligible g_x and g_y values. The g_z value for Dy^{III} is close to the expected $g_z = 20$ for a pure $M_J = |\pm 15/2>$ ground state. This is confirmed by the calculated composition of $M_J = 0.94 \ |\pm 15/2> + 0.06 \ |\pm 11/2>$ for the ground doublet state of **1** (see Table S5 for the wavefunction composition). The calculated ground-state easy axis (Figure 6) for the Dy^{III} ion is oriented perpendicular to the plane formed by the 2,2′-bipyridine moieties as expected for an oblate ion with this coordination sphere [31,36].

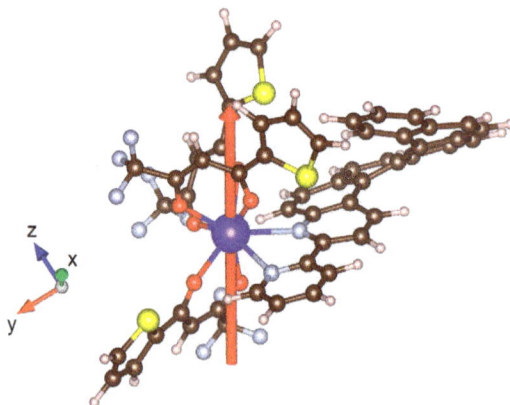

Figure 6. Representation of complex **1** with the theoretical orientation of the easy magnetic axis of the Dy^{III} center.

Even if the uniaxiality of the anisotropy is not strictly associated to slow relaxation of the magnetization [46], in most of the cases reported in the literature for Dy^{III} this assumption is valid.

Indeed, in this case, the magnetic relaxation pathways can also be easily interpreted on the basis of magnetic transition moments (Figure 7) calculated with the SINGLE_ANISO program [47,48]. It has to be pointed out that in the latter not all the contributions are included. Indeed, the coupling of spin-phonon degrees of freedom in the SMM relaxation is not taken into account in the ab initio model whereas it has been recently evidenced of general importance [49,50]. However, these discrepancies are common in literature [3,51,52] and the magnetic transition moments calculated in this work still leads to a fairly good qualitative picture. Indeed, no direct transition between the two M_J states of the ground doublet or Orbach processes from the ground state are expected whereas relaxation mechanisms involving states from the third M_J state are highly probable. A non-negligible Orbach process has been also found between the second and third M_J states. The calculations indicate a difference between the calculated energy barrier ($\Delta = 82$ cm^{-1}) and the experimental barrier ($\Delta = 39$–68 cm^{-1}). However, the discrepancy between these values can be, reasonably, ascribed in the spin-phonon contributions mentioned above.

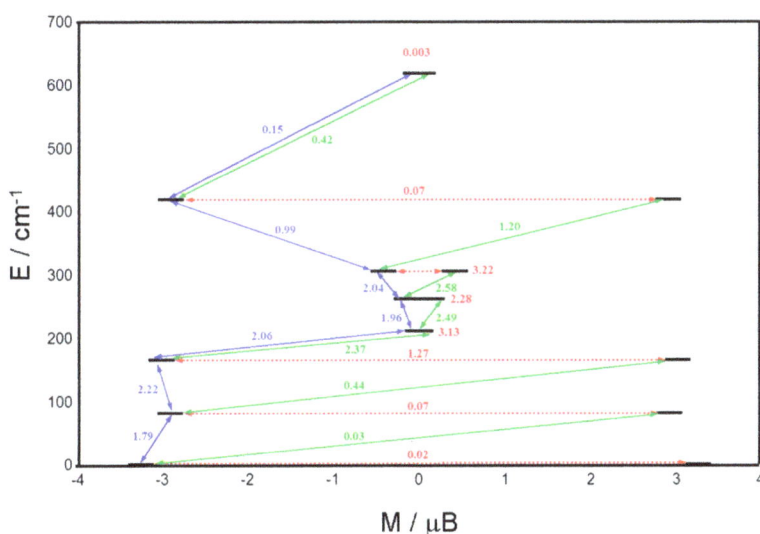

Figure 7. Computed magnetization blocking barrier in complex **1** for the Dy^{III} ion. Numbers provided on each arrow are the mean absolute values for the corresponding matrix elements of the magnetic transition dipole moment.

3. Materials and Methods

3.1. Synthesis. General Procedures and Materials

The precursor Dy(tta)$_3$·2H$_2$O [53] (tta$^-$ = 2-thenoyltrifluoroacetonate anion) and the ligand 3-(2-pyridyl)-4-aza[6]-helicene [40,54] **L** were synthesized following previously reported methods. All other reagents were commercially available and used without further purification.

3.2. Synthesis of Complex [Dy(tta)$_3$(**L**)]·CH$_2$Cl$_2$·C$_7$H$_8$ (1)

Dy(tta)$_3$·2H$_2$O (17.2 mg, 0.02 mmol) were dissolved in 5 mL of CH$_2$Cl$_2$ and then added to a solution of 5 mL of CH$_2$Cl$_2$ containing 8.3 mg of **L** (0.02 mmol). After 20 min of stirring, 30 mL of toluene were layered at 4 °C in the dark. Slow diffusion following by slow evaporation lead to yellow single crystals which are suitable for X-ray studies. Yield: 15.2 mg (54% based on Dy). Anal. Calcd (%) for C$_{62}$H$_{40}$Cl$_2$DyF$_9$N$_2$O$_6$S$_3$: C 52.78, H 2.84, N 1.99; found: C 52.72, H 2.81 N, 2.09. I.R. 3426 (m),

2923 (w), 1604 (s), 1588 (m), 1566 (m), 1493 (m), 1468 (m), 1429 (m), 1256 (s), 117 (m), 1091 (w), 1046 (w), 991 (w), 963 (w), 848 (w), 815 (w), 797 (s), 773 (m), 755 (m), 654 (m), 542 (m), 503 (m) cm^{-1}.

3.3. Crystallography

Single crystal of [Dy(tta)$_3$(**L**)]·CH$_2$Cl$_2$·C$_7$H$_8$ (**1**) was mounted on a APEXII Bruker-AXS diffractometer for data collection (MoKα radiation source, λ = 0.71073 Å). The structure was solved by direct methods using the SIR-97 program and refined with a full matrix least-squares method on F^2 using the SHELXL-97 program [55,56]. Crystallographic data are summarized in Table S1. Complete crystal structure results as a CIF file including bond lengths, angles, and atomic coordinates are deposited as Supporting Information. CCDC number is 1510324 for compound **1**.

3.4. Physical Measurements

The elementary analyses of the compound were performed at the Centre Régional de Mesures Physiques de l'Ouest, Rennes, France. The dc magnetic susceptibility measurements were performed on solid polycrystalline sample with a Quantum Design MPMS-XL SQUID magnetometer between 2 and 300 K in an applied magnetic field of 0.02 T for temperatures in the range 2–20 K, 0.2 T between 20 and 80 K and 1 T for temperatures between 80 and 300 K. These measurements were all corrected for the diamagnetic contribution as calculated with Pascal's constants. The ultra-low temperature measurements (below 1.8 K) were performed with the help of a ^{3}He insert.

3.5. Computational Details

Wavefunction-based calculations were carried out on molecular structures of **1** by using the SA-CASSCF/RASSI-SO approach, as implemented in the MOLCAS quantum chemistry package (versions 8.0) [57]. In this approach, the relativistic effects are treated in two steps on the basis of the Douglas–Kroll Hamiltonian. First, the scalar terms were included in the basis-set generation and were used to determine the spin-free wavefunctions and energies in the complete active space self-consistent field (CASSCF) method [58]. Next, spin-orbit coupling was added within the restricted-active-space-state-interaction (RASSI-SO) method, which uses the spin-free wavefunctions as basis states [59,60]. The resulting wavefunctions and energies are used to compute the magnetic properties and g-tensors of the lowest states from the energy spectrum by using the pseudo-spin $S = 1/2$ formalism in the SINGLE-ANISO routine [47,48]. Cholesky decomposition of the bielectronic integrals was employed to save disk space and speed-up the calculations [61]. For **1** the active space of the self consistent-field (CASSCF) method consisted of the nine 4f electrons of the Dy^{III} ion spanning the seven 4f orbitals, i.e., CAS(9,7)SCF. State-averaged CASSCF calculations were performed for all of the sextets (21 roots), all of the quadruplets (224 roots), and 300 out of the 490 doublets (due to hardware limitations) of the Dy^{III} ion. Twenty-one sextets, 128 quadruplets, and 107 doublets were mixed through spin–orbit coupling in RASSI-SO. All atoms were described by ANO-RCC basis sets [62–64]. The following contractions were used: [8s7p4d3f2g1h] for Dy, [4s3p2d1f] for the O and N atoms, [3s2p1d] for the C and F atoms, [4s3p1d] for the S atoms and [2s1p] for the H atoms. The atomic positions were extracted from the X-ray crystal structures. Only the position of the H and F atoms were optimized on the Y^{III} parent complexes with the Gaussian 09 (revision D.01) package [65] employing the PBE0 hybrid functional [66,67]. The "Stuttgart/Dresden" basis sets [68] and effective core potentials were used to describe the yttrium atom, whereas all other atoms were described with the SVP basis sets [69].

4. Conclusions

In the course of intermixing chirality offered by the nature of the ligands and SMM properties, we extend herein the family of [6]-helicene-based lanthanide SMM [27]. We report the synthesis of the complex [Dy(**L**)(tta)$_3$] with **L** the chiral 3-(2-pyridyl)-4-aza[6]-helicene ligand (tta$^-$ = 2-thenoyltrifluoroaacetonate). Its racemic form was structurally and magnetically characterized.

[Dy(**L**)(tta)$_3$] behaves as a single molecule magnet in its crystalline phase. As expected, the substitution of hfac$^-$ ligands by tta$^-$ moieties enhances the magnetic behavior with the opening of an hysteresis loop at 0.50 K that was only observed for the enantiopure forms in the case of [Dy(**L**)(hfac)$_3$] [27]. The electronic structure of the complex has been elucidated by mean of SA-CASSCF/RASSI-SO calculations, highlighting the nature of the ground state and contributing in the interpretation of experimental evidences. A qualitative picture of the magnetization blocking barrier is also reported. In the near future, we will pursue our investigation of chiral lanthanide-based SMMs that may offer new perspectives in both the domains of molecular magnetism and chirality with the potential access of properties such as circularly polarized luminescence (CPL) activity that remains anecdotic for lanthanide compounds to date.

Supplementary Materials: The following are available online at www.mdpi.com/2312-7481/3/1/2/s1, Figure S1: ORTEP view of **1**. Thermal ellipsoids are drawn at 30% probability. Hydrogen atoms and solvent molecules of crystallization are omitted for clarity., Figure S2: Frequency dependence of the ac susceptibility components χ_M' and χ_M'' at 10 K and in zero external dc field for compound **1** with the best fitted curve with extended Debye model, Figure S3: Frequency dependence of the ac susceptibility components χ_M' and χ_M'' at 10 K and in 1000 Oe external dc field for compound **1** with the best fitted curve with extended Debye model; Figure S4: Cole–Cole plots using the ac data performed under zero magnetic field. The black lines correspond to the fit with a generalized Debye model; Figure S5: Cole–Cole plots using the ac data performed under 1000 Oe magnetic field. The black lines correspond to the fit with a generalized Debye model; Figure S6: Magnetic hysteresis loop of **1** measured at 0.5, 1.0 and 1.5 K; Table S1: X-ray crystallographic data of **1**; Table S2: SHAPE analysis for **1**; Table S3: Best fitted parameters (χ_T, χ_S, τ and α) with the extended Debye model **1** at zero field in the temperature range 2.0–15 K; Table S4: Best fitted parameters (χ_T, χ_S, τ and α) with the extended Debye model **1** at 1 kOe in the temperature range 1.8–5 K; Table S5: Computed energies, *g*-tensor and wavefunction composition of the ground state doublet in the effective spin $\frac{1}{2}$ model for **1**.

Acknowledgments: This work was supported by Région Bretagne, Rennes Métropole, CNRS, Université de Rennes 1. G.F.G gratefully acknowledges the European Commission through the ERC-AdG 267746 MolNanoMas (project N. 267746) and the ANR (ANR-13-BS07-0022-01) for financial support. J.C. with N.S. and J.-K.O.-Y., respectively, thank the ANR (ANR-10-BLAN-724-1-NCPCHEM) and the Chinese Scholarship Council for financial support. B.L.G. and G.F.G. thank the French GENCI/IDRIS-CINES center for high-performance computing resources.

Author Contributions: J.-K.O.-Y., and N.S. performed the organic syntheses; F.P. performed the coordination chemistry, crystallizations, the single crystal X-ray diffraction experiments and structure refinements; O.C. performed the magnetic measurements; J.F.G. analyzed the magnetic measurements; G.F.-G., F.T., and B.L.G. performed the ab initio calculations; J.C., and L.O. discussed the results and commented on the manuscript; and F.P., O.C., and B.L.G. conceived and designed the experiments and contributed equally to the writing of the article.

Abbreviations

The following abbreviations are used in this manuscript:

SMM	Single Molecule Magnet
TTF	TetraThiaFulvalene
CH$_2$Cl$_2$	Dichloromethane
hfac	1,1,1,5,5,5-hexafluoroacetylacetonate
tta	2-thenoyltrifluoroacetonate
PCM	Polarizable Continuum Model
CASSCF	Complete Active Space Self-Consistent Field
RASSI-SO	Restricted Active Space State Interaction—Spin-Orbit

References

1. Sessoli, R.; Powell, A.K. Strategies towards single molecule magnets based on lanthanide ions. *Coord. Chem. Rev.* **2009**, *253*, 2328–2341. [CrossRef]

2. Liddle, S.T.; van Slageren, J. Improving f-element single molecule magnets. *Chem. Soc. Rev.* **2015**, *44*, 6655–6669. [CrossRef] [PubMed]

3. Pedersen, K.S.; Dreiser, J.; Weihe, H.; Sibille, R.; Johannesen, H.V.; Sorensen, M.A.; Nielsen, B.E.; Sigrist, M.; Mutka, H.; Rols, S.; et al. Design of Single-Molecule Magnets: Insufficiency of the Anisotropy Barrier as the Sole Criterion. *Inorg. Chem.* **2015**, *54*, 7600–7606. [CrossRef] [PubMed]

4. Pedersen, K.S.; Ariciu, A.; McAdams, S.; Weihe, H.; Bendix, J.; Tuna, F.; Piligkos, S. Toward Molecular 4f Single-Ion Magnet Qubits. *J. Am. Chem. Soc.* **2016**, *138*, 5801–5804. [CrossRef] [PubMed]

5. Mannini, M.; Pineider, F.; Sainctavit, P.; Danieli, C.; Otero, E.; Sciancalepore, C.; Talarico, A.M.; Arrio, M.-A.; Cornia, A.; Gatteschi, D. Magnetic memory of a single-molecule quantum magnet wired to a gold surface. *Nat. Mater.* **2009**, *8*, 194–197. [CrossRef] [PubMed]

6. Rocha, A.R.; García-Suárez, V.M.; Bailey, S.W.; Lambert, C.J.; Ferrer, J.; Sanvito, S. Towards Molecular Spintronics. *Nat. Mater.* **2005**, *4*, 335–339. [CrossRef] [PubMed]

7. Rinehart, J.D.; Long, J.R. Exploiting single-ion anisotropy in the design of f-element single-molecule magnets. *Chem. Sci.* **2011**, *2*, 2078–2085. [CrossRef]

8. Lucaccini, E.; Briganti, M.; Perfetti, M.; Vendier, L.; Costes, J.-P.; Totti, F.; Sessoli, R.; Sorace, L. Relaxation Dynamics and Magnetic Anisotropy in a Low-Symmetry Dy^{III} Complex. *Chem. Eur. J.* **2016**, *22*, 5552–5562. [CrossRef] [PubMed]

9. Zhang, P.; Zhang, L.; Tang, J. Lanthanide single molecule magnets: Progress and perspective. *Dalton. Trans.* **2015**, *44*, 3923–3929. [CrossRef] [PubMed]

10. Zhang, P.; Jung, J.; Zhang, L.; Tang, J.; Le Guennic, B. Elucidating the Magnetic Anisotropy and Relaxation Dynamics of Low-Coordinate Lanthanide Compounds. *Inorg. Chem.* **2016**, *55*, 1905–1911. [CrossRef] [PubMed]

11. Cucinotta, G.; Perfetti, M.; Luzon, J.; Etienne, M.; Car, P.E.; Caneschi, A.; Calvez, G.; Bernot, K.; Sessoli, R. Magnetic Anisotropy in a Dysprosium/DOTA Single-Molecule Magnet: Beyond Simple Magneto-Structural Correlations. *Angew. Chem. Int. Ed.* **2012**, *51*, 1606–1610. [CrossRef] [PubMed]

12. Pointillart, F.; Le Guennic, B.; Cador, O.; Maury, O.; Ouahab, L. Lanthanide and Ion and Tetrathiafulvalene-Based Ligand as a "Magic" Couple toward Luminescence, Single Molecule Magnets, and Magnetostructural Correlations. *Acc. Chem. Res.* **2015**, *48*, 2834–2842. [CrossRef] [PubMed]

13. Pointillart, F.; Jung, J.; Berraud-Pache, R.; Le Guennic, B.; Dorcet, V.; Golhen, S.; Cador, O.; Maury, O.; Guyot, Y.; Decurtins, S.; et al. Luminescence and Single-Molecule Magnet Behavior in Lanthanide Complexes Involving a Tetrathiafulvalene-Fused Dipyridophenazine Ligand. *Inorg. Chem.* **2015**, *54*, 5384–5397. [CrossRef] [PubMed]

14. Long, J.; Vallat, R.; Ferreira, R.A.S.; Carlos, L.D.; Almeida Paz, F.A.; Guari, Y.; Larionova, J. A bifunctional luminescent single-ion magnet: Towards correction between luminescence studies and magnetic slow relaxation processes. *Chem. Commun.* **2012**, *48*, 9974–9976. [CrossRef] [PubMed]

15. Pointillart, F.; Le Guennic, B.; Golhen, S.; Cador, O.; Maury, O.; Ouahab, L. A redox active luminescent ytterbium single-molecule magnet. *Chem. Commun.* **2013**, *49*, 615–617. [CrossRef] [PubMed]

16. Sabbatini, N.; Guardigli, M.; Manet, I. *Handbook of the Physics and Chemistry of Rare Earths*; Elsevier: Amsterdam, The Netherlands, 1996; Volume 23, p. 69.

17. Comby, S.; Bünzli, J.-C.G. *Handbook on the Physics and Chemistry of Rare Earths*; Elsevier: Amsterdam, The Netherlands, 2007; Volume 37, Chapter 235.

18. Parker, D. Luminescent Lanthanide Sensors for pH, pO_2 and Selected Anions. *Coord. Chem. Rev.* **2000**, *205*, 109–130. [CrossRef]

19. Parker, D. Excitement in f block: Structure, dynamics and function of nine-coordinate chiral lanthanide complexes in aqueous media. *Chem. Soc. Rev.* **2004**, *33*, 156–165. [CrossRef] [PubMed]

20. Bünzli, J.-C.G.; Piguet, C. Taking advantage of luminescent lanthanide ions. *Chem. Soc. Rev.* **2005**, *34*, 1048–1077. [CrossRef] [PubMed]

21. Eliseeva, S.V.; Bünzli, J.-C.G. Lanthanide luminescence for functional materials and bio-sciences. *Chem. Soc. Rev.* **2010**, *39*, 189–227. [CrossRef] [PubMed]

22. D'Aléo, A.; Pointillart, F.; Ouahab, L.; Andraud, C.; Maury, O. Charge transfer excited states sensitization of lanthanide emitting from the visible to the near-infra-red. *Coord. Chem. Rev.* **2012**, *256*, 1604–1620. [CrossRef]

23. VanVleck, J.H. The Puzzle of rare-earth Spectra in Solids. *J. Phys. Chem.* **1937**, *41*, 67–80. [CrossRef]

24. Wang, X.; Chang, H.; Xie, J.; Zhao, B.; Liu, B.; Xu, S.; Pei, W.; Ren, N.; Huang, L.; Huang, W. Recent developments in Lanthanide-based luminescent probes. *Coord. Chem. Rev.* **2014**, *273–274*, 201–212. [CrossRef]

25. Train, C.; Ghoerghe, R.; Krstic, V.; Chamoreau, L.; Ovanesyan, N.S.; Rikken, G.L.J.A.; Gruselle, M.; Verdaguer, M. Strong magneto-chiral dichroism in enantiopure chiral ferromagnets. *Nat. Mater.* **2008**, *7*, 729–734. [CrossRef] [PubMed]

26. Sessoli, R.; Boulon, M.-E.; Caneschi, A.; Mannini, M.; Poggini, L.; Wilhelm, F.; Rogalev, A. Strong magneto-chiral dichroism in a paramagnetic molecular helix observed by hard X-rays. *Nat. Phys.* **2014**, *11*, 69–74. [CrossRef] [PubMed]

27. Ou-Yang, J.-K.; Saleh, N.; Fernandez Garcia, G.; Norel, L.; Pointillart, F.; Guizouarn, T.; Cador, O.; Totti, F.; Ouahab, L.; Crassous, J.; et al. Improved Slow Magnetic Relaxation in Optically Pure Helicene-Based Dy^{III} Single Molecule Magnet. *Chem. Commun.* **2016**, *52*, 14474–14477. [CrossRef] [PubMed]

28. Wang, Y.; Li, X.L.; Wang, T.W.; Song, Y.; You, X.Z. Slow Relaxation Processes and Single-Ion Magnetic Behaviors in Dysprosium-Containing Complexes. *Inorg. Chem.* **2010**, *49*, 969–976. [CrossRef] [PubMed]

29. Li, D.-P.; Wang, T.-W.; Li, C.-H.; Liu, D.-S.; Li, Y.-Z.; You, X.-Z. Single-ion magnets based on mononuclear lanthanide complexes with chiral Schiff base ligands [Ln(FTA)$_3$L] (Ln = Sm, Eu, Gd, Tb and Dy). *Chem. Commun.* **2010**, *46*, 2929–2931. [CrossRef] [PubMed]

30. Norel, L.; Bernot, K.; Feng, M.; Roisnel, T.; Caneschi, A.; Sessoli, R.; Rigaut, S. A carbon-rich ruthenium decorated dysprosium single molecule magnet. *Chem. Commun.* **2012**, *48*, 3948–3950. [CrossRef] [PubMed]

31. Bosson, J.; Gouin, J.; Lacour, J. Cationic triangulenes and helicenes: Synthesis, chemical stability, optical properties and extended applications of these unusual dyes. *Chem. Soc. Rev.* **2014**, *43*, 2824–2840. [CrossRef] [PubMed]

32. Saleh, N.; Shen, C.; Crassous, J. Helicene-based transition metal complexes: Synthesis, properties and applications. *Chem. Sci.* **2014**, *5*, 3680–3694. [CrossRef]

33. Storch, J.; Zadny, J.; Strasak, T.; Kubala, M.; Syroka, J.; Dusek, M.; Cirkva, V.; Matejka, P.; Krbal, M.; Vacek, J. Synthesis and Characterization of a Helicene-Based Imidazolium Salt and its Application in Organic Molecular Electronics. *Chem. Eur. J.* **2014**, *21*, 2343–2347. [CrossRef] [PubMed]

34. Mendola, D.; Saleh, N.; Hellou, N.; Vanthuyne, N.; Roussel, C.; Toupet, L.; Castiglione, F.; Melone, F.; Caronna, T.; Fontana, F.; et al. Synthesis and Structural Properties of Aza[n]helicene Platinum Complex: Control of Cis and Trans Stereochemistry. *Inorg. Chem.* **2016**, *55*, 2009–2017. [CrossRef] [PubMed]

35. Vacek, J.; Vacek Chocholousova, J.; Stara, I.G.; Stary, I.; Dubi, Y. Mechanical tuning of conductance and thermopower in helicene molecular junctions. *Nanoscale* **2015**, *7*, 8793–8802. [CrossRef] [PubMed]

36. Jung, J.; da Cunha, T.T.; Le Guennic, B.; Pointillart, F.; Pereira, C.L.M.; Luzon, J.; Golhen, S.; Cador, O.; Maury, O.; Ouahab, L. Magnetic Studies of Redox-Active Tetrathiafulvalene-Based Complexes: Dysprosium vs. Ytterbium Ananlogues. *Eur. J. Inorg. Chem.* **2014**, *2014*, 3888–3894. [CrossRef]

37. Cosquer, G.; Pointillart, F.; Golhen, S.; Cador, O.; Ouahab, L. Slow Magnetic Relaxation in Condensed versus Dispersed Dysprosium(III) Mononuclear Complexes. *Chem. Eur. J.* **2013**, *19*, 7895–7905. [CrossRef] [PubMed]

38. Da Cunha, T.T.; Jung, J.; Boulon, M.-E.; Campo, G.; Pointillart, F.; Pereira, C.L.M.; Le Guennic, B.; Cador, O.; Bernot, K.; Pineider, F.; et al. Magnetic Poles Determinations and Robustness of Memory Effect upon Solubilization in a Dy^{III}-Based Single Ion Magnet. *J. Am. Chem. Soc.* **2013**, *135*, 16332–16335. [CrossRef] [PubMed]

39. Jiménez, J.R.; Díaz-Ortega, I.F.; Ruiz, E.; Aravena, D.; Pope, S.J.A.; Colacio, E.; Herrera, J.M. Lanthanide tetrazolate complexes combining Single-Molecule Magnet and luminescence properties: The effect of the replacement of tetrazolate N3 by β-diketonate ligands on the anisotropy energy barrier. *Chem. Eur. J.* **2016**, *22*, 14548–14559. [CrossRef] [PubMed]

40. Saleh, N.; Moore, B., II; Srebro, M.; Vanthuyne, N.; Toupet, L.; Williams, J.A.G.; Roussel, C.; Deol, K.K.; Muller, G.; Autschbach, J.; et al. Acid/Base-Triggered Switching of Circularly Polarized Luminescence and Electronic Circular Dichroism in Organic and Organometallic Helicenes. *Chem. Eur. J.* **2015**, *21*, 1673–1681. [CrossRef] [PubMed]

41. Llunell, M.; Casanova, D.; Cirera, J.; Bofill, J.M.; Alemany, P.; Alvarez, S.S. *SHAPE*, version 2.1; University of Barcelona: Barcelona, Spain, 2013.

42. Kahn, O. *Molecular Magnetism*; VCH: Weinhem, Germany, 1993.

43. Dekker, C.; Arts, A.F.M.; Wijn, H.W.; van Duyneveldt, A.J.; Mydosh, J.A. Activated dynamics in a two-dimensional Ising spin glass: Rb$_2$Cu$_{1-x}$Co$_x$F$_4$. *Phys. Rev. B* **1989**, *40*, 11243–11251. [CrossRef]

44. Cole, K.S.; Cole, R.H. Dipersion and Absorption in Dielectrics I. Alternating Current Charcateristics. *J. Chem. Phys.* **1941**, *9*, 341–351. [CrossRef]

45. Baldovi, J.J.; Duan, Y.; Morales, R.; Gaita-Ariño, A.; Ruiz, E.; Coronado, E. Rational design of lanthanoid Single-Ion Magnets: Predictive power of the theoretical models. *Chem. Eur. J.* **2016**, *22*, 13532–13539. [CrossRef] [PubMed]

46. Lucaccini, E.; Sorace, L.; Perfetti, M.; Costes, J.-P.; Sessoli, R. Beyond the anisotropy barrier: Slow relaxation of the magnetization in both easy-axis and easy-plane Ln(trensal) complexes. *Chem. Commun.* **2014**, *50*, 1648–1651. [CrossRef] [PubMed]

47. Chibotaru, L.F.; Ungur, L. Ab initio calculation of anisotropic magnetic properties of complexes. I. Unique definition of pseudospin Hamiltonians and their derivation. *J. Chem. Phys.* **2012**, *137*, 064112–064122. [CrossRef] [PubMed]

48. Chibotaru, L.F.; Ungur, L.; Soncini, A. The Origin of Nonmagnetic Kramers Doublets in the Ground State of Dysprosium Triangles: Evidence for a Toroidal Magnetic Moment. *Angew. Chem. Int. Ed.* **2008**, *47*, 4126–4129. [CrossRef] [PubMed]

49. Lunghi, A.; Totti, F. The role of Anisotropic Exchange in Single Molecule Magnets: A CASSCF/NEVPT2 Study of the Fe_4 SMM Building Block $[Fe_2(OCH_3)_2(dbm)_4]$ Dimer. *Inorganics* **2016**, *4*, 28–38. [CrossRef]

50. Tesi, L.; Lunghi, A.; Atzori, M.; Lucaccini, E.; Sorace, L.; Totti, F.; Sessoli, R. Giant spin-phonon bottleneck effects in evaporable vanadyl-based molecules with long spin coherence. *Dalton Trans.* **2016**, *45*, 16635–16643. [CrossRef] [PubMed]

51. Zadrozny, J.M.; Long, J.R. Slow Magnetic Relaxation at Zero Field in the tetrahedral Complex $[Co(SPh)_4]^{2-}$. *J. Am. Chem. Soc.* **2011**, *133*, 20732–20734. [CrossRef] [PubMed]

52. Freedman, D.E.; Harman, W.H.; Harris, T.D.; Long, G.H.; Chang, C.J.; Long, J.R. Slow Magnetic Relaxation in a High-Spin Iron(II) Complex. *J. Am. Chem. Soc.* **2010**, *132*, 1224–1225. [CrossRef] [PubMed]

53. Vooshin, A.I.; Shavaleev, N.M.; Kazakov, V.P. Chemiluminescence of praseodymium (III), neodymium (III) and ytterbium (III) β-diketonates in solution excited from 1,2-dioxetane decomposition and singlet-singlet energy transfer from ketone to rare-earth β-diketonates. *J. Lumin.* **2000**, *91*, 49–58. [CrossRef]

54. Saleh, N.; Srebro, M.; Reynaldo, T.; Vanthuyne, N.; Toupet, L.; Chang, V.Y.; Muller, G.; Williams, J.A.G.; Roussel, C.; Autschbach, J.; et al. *Enantio*-Enriched CPL-active helicene-bipyridine-rhenium complexes. *Chem. Commun.* **2015**, *51*, 3754–3757. [CrossRef] [PubMed]

55. Sheldrick, G.M. *SHELX97—Programs for Crystal Structure Analysis (Release 97-2)*; Institüt für Anorganische Chemie der Universität: Göttingen, Germany, 1998.

56. Altomare, A.; Burla, M.C.; Camalli, M.; Cascarano, G.L.; Giacovazzo, C.; Guagliardi, A.; Moliterni, A.G.G.; Polidori, G.; Spagna, R. SIR97: A new tool for crystal structure determination and refinement. *J. Appl. Cryst.* **1999**, *32*, 115–119. [CrossRef]

57. Aquilante, F.; de Vico, L.; Ferré, N.; Ghigo, G.; Malmqvist, P.A.; Neogrady, P.; Bondo Pedersen, T.; Pitonak, M.; Reiher, M.; Roos, B.O.; et al. MOLCAS 7: The Next Generation. *J. Comput. Chem.* **2010**, *31*, 224–247. [CrossRef] [PubMed]

58. Roos, B.O.; Taylor, P.R.; Siegbahn, P.E.M. A complete active space SCF method (CASSCF) using a density matrix formulated super-CI approach. *Chem. Phys.* **1980**, *48*, 157–288. [CrossRef]

59. Malmqvist, P.A.; Roos, B.O.; Schimmelpfennig, B. The restricted active space (RAS) state interaction approach with spin-orbit coupling. *Chem. Phys. Lett.* **2002**, *357*, 230–240. [CrossRef]

60. Malmqvist, P.A.; Roos, B.O. The CASSCF state interaction method. *Chem. Phys. Lett.* **1989**, *155*, 189–194. [CrossRef]

61. Aquilante, F.; Malmqvist, P.-A.; Pedersen, T.-B.; Ghosh, A.; Roos, B.O. Decomposition-Based Multiconfiguration Second-Order Perturbation Theory (CD-CASPT2): Application to the Spin-State Energetics of Co^{III}(diiminato)(NPh). *J. Chem. Theory Comput.* **2008**, *4*, 694–702. [CrossRef] [PubMed]

62. Roos, B.O.; Lindh, R.; Malmqvist, P.-A.; Veryazov, V.; Widmark, P.-O. Main Group Atoms and Dimers Studied with a New Relativistic ANO Basis Set. *J. Phys. Chem. A* **2004**, *108*, 2851–2858. [CrossRef]

63. Roos, B.O.; Lindh, R.; Malmqvist, P.-A.; Veryazov, V.; Widmark, P.-O. New Relativistic ANO Basis Sets for Transition Metal Atoms. *J. Phys. Chem. A* **2005**, *109*, 6576–6586. [CrossRef] [PubMed]

64. Roos, B.O.; Lindh, R.; Malmqvist, P.-A.; Veryazov, V.; Widmark, P.-O.; Borin, A.-C. New relavistic Atomic Natural Orbital Basis Sets for lanthanide Atoms with Applications to the Ce Diatom and LuF_3. *J. Phys. Chem. A* **2008**, *112*, 11431–11435. [CrossRef] [PubMed]

65. Frisch, M.J.; Trucks, G.W.; Schlegel, H.B.; Scuseria, G.E.; Robb, M.A.; Cheeseman, J.R.; Scalmani, G.; Barone, V.; Mennucci, B.; Petersson, G.A.; et al. *Gaussian 09, Revision A.02*; Gaussian Inc.: Wallingford, CT, USA, 2009.
66. Perdew, J.P.; Burke, K.; Ernzerhof, M. Generalized Gradient Approximation Made Simple. *Phys. Rev. Lett.* **1996**, *77*, 3865–3868. [CrossRef] [PubMed]
67. Adamo, C.; Barone, V. Toward reliable density functional methods without adjustable parameters: The PBE0 model. *J. Chem. Phys.* **1999**, *110*, 6158–6170. [CrossRef]
68. Dolg, M.; Stoll, H.; Preuss, H. A combination of quasirelativistic pseudopotential and ligand field calculations for lanthanoid compounds. *Theor. Chim. Acta* **1993**, *85*, 441–450. [CrossRef]
69. Weigend, F.; Ahlrichs, R. Balanced basis sets of split valence, triple zeta valence and quadruple zeta valence quality for H to Rn: Design and assessment of accuracy. *Phys. Chem. Chem. Phys.* **2005**, *7*, 3297–3305. [CrossRef] [PubMed]

magnetochemistry

MDPI

Article

Slow Magnetic Relaxation of Lanthanide(III) Complexes with a Helical Ligand

Hisami Wada [1], Sayaka Ooka [1], Daichi Iwasawa [2], Miki Hasegawa [2,*] and Takashi Kajiwara [1,*]

[1] Department of Chemistry, Faculty of Science, Nara Women's University, Kita-uoya Nishi-machi, Nara 630-8506, Japan; samph.283a.27@gmail.com (H.W.); oas_ooka@yahoo.co.jp (S.O.)

[2] Department of Chemistry and Biological Science, College of Science and Engineering, Aoyama Gakuin University, Fuchinobe, Chuo-ku, Sagamihara, Kanagawa 252-5258, Japan; asenal0123456789@gmail.com

*   Correspondence: hasemiki@chem.aoyama.ac.jp (M.H.); kajiwara@cc.nara-wu.ac.jp (T.K.); Tel.: +81-42-759-6221 (M.H.); +81-742-20-3402 (T.K.)

Academic Editor: Kevin Bernot
Received: 31 October 2016; Accepted: 30 November 2016; Published: 8 December 2016

Abstract: Isostructural Ln(III) mononuclear complexes $[Ln(NO_3)_2L]PF_6 \cdot MeCN$ (Ln = Nd, Tb, or Dy; L denotes a helical hexa-dentate ligand) were synthesized, and their slow magnetic relaxation behavior was investigated. In these complexes, oblate-type Ln(III) ions are located in an axially stressed ligand field with two nitrate anions, and can exhibit single-molecule magnet (SMM) behavior. Field-induced SMM behavior was observed for Nd(III) and Dy(III) complexes under an applied bias DC field of 1000 Oe.

Keywords: lanthanide complex; slow magnetic relaxation; single-molecule magnet; crystal structure; AC susceptibility

1. Introduction

Single-molecule magnets (SMMs) are fascinating molecule-based nanomaterials [1–8]. They are characterized by slow magnetization at low temperature, of which the magnetic anisotropy plays an essential role in preventing the flipping of the magnetic moment [1]. As the orbital angular momentum of 4f electrons are maintained and unquenched, each lanthanide(III) (Ln(III)) ion possesses a large magnetic moment correlated with the total angular momentum, J, which is defined by the length of the resultant vector of the spin angular momentum, S, and the orbital angular momentum, L. The expected J values of Ce(III) ($L = 3$, $S = 1/2$), Nd(III) ($L = 6$, $S = 3/2$), Tb(III) ($L = 3$, $S = 3$), and Dy(III) ($L = 5$, $S = 5/2$) are $5/2$, $9/2$, 6, and $15/2$, respectively. In each Ln(III) ion, the contribution of the orbital angular momentum to the total angular momentum is greater than that of the spin angular momentum. Hence, each sub-level, which is characterized by magnetic quantum number J_z, has a distinguishable electronic distribution and interacts with the crystal field in a different manner. In these Ln(III) ions, sub-levels with the highest $|J_z|$ quantum numbers, that is, $J_z = \pm 5/2$ for Ce(III), $\pm 9/2$ for Nd(III), ± 6 for Tb(III), and $\pm 15/2$ for Dy(III), have oblate-shaped electronic distributions and are called oblate-type Ln(III) ions [9–11]. These J_z sublevels are relatively stabilized when the Ln(III) ion is located in an axially stressed crystal field, where the equatorial electronic repulsion is smaller than the axial electronic repulsion, since other J_z sub-levels are more destabilized under these conditions. This situation can be realized in complexes by combining neutral multi-dentate ligands, which occupy the equatorial positions of the Ln(III) ion and anionic ligands, which occupy the axial positions [11]. In a previous paper, we reported the syntheses and crystal structures of a series of Ln(III) complexes constructed from a single helical ligand (L, Figure 1) and two nitrate anions, $[LnL(NO_3)_2]PF_6 \cdot MeCN$ (Ln = Nd, Eu, Gd, and Tb) and $[HoL(NO_3)_2]PF_6 \cdot 2MeCN$, with the aim of highly efficient f-f luminescence by UV irradiation of aromatic bipyridine moieties in L [12]. This series of complexes was beneficial to study the magneto-structural correlation of slow magnetic relaxation phenomena, not only because the coordination structure was beneficial to the magnetic anisotropy design, but also because similar molecular structures were achieved in a wide variety of Ln(III) ions containing both light and heavy lanthanide ions. Several years ago, we reported that the slow magnetic relaxation of light lanthanide complexes can be achieved using a synthetic strategy similar to that employed for the heavy lanthanide ions [13–15]. However, examples of light lanthanide complexes that show slow magnetic relaxation phenomena are somewhat rare [13–18]. In this study, we have newly prepared a Dy(III) complex with crystal structure confirmed to be isostructural with those of the reported Nd(III) and Tb(III) complexes, and investigated the slow magnetic relaxation phenomena of this complex family, [LnL(NO$_3$)$_2$]PF$_6$·MeCN (Ln = Nd (**1**), Tb (**2**), or Dy(**3**)), each of which contains oblate-type lanthanide ions as a magnetic center.

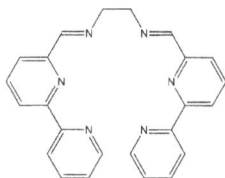

Figure 1. Structure of ligand L.

2. Results

2.1. Synthesis and Characterization

The syntheses and crystal structures of Nd(III), Eu(III), Gd(III), Tb(III), and Ho(III) complexes have already been reported with the aim of the investigating their optical properties [12]. The mono-nuclear complexes of Nd(III), Eu(III), Gd(III), and Tb(III) were found to be crystallographically isostructural, while the Ho(III) complex was different. The solvent molecule content in the crystals was also slightly different between the Ho(III) complex and the other complexes, with the former formulated as [HoL(NO$_3$)$_2$]PF$_6$·2MeCN and the latter as [LnL(NO$_3$)$_2$]PF$_6$·MeCN. These differences may be attributed to lanthanide contraction. Gradual changes in the ionic radii of the lanthanide ions sometimes cause a clear change in the crystal structures when the lanthanide ions varied from lighter to heavier elements. In the aforementioned series, the Dy(III) complex may be located on the boundary of the structural change of the crystals. X-ray analysis revealed that the Dy(III) complex was isomorphous with the Tb(III) complex and formulated as [DyL(NO$_3$)$_2$]PF$_6$·MeCN. As we could handle this series of isomorphous complexes, which include three oblate-type lanthanide complexes, we decided to investigate the magnetism of these complexes, which include Nd(III), Tb(III), and Dy(III), as magnetic centers.

We obtained the newly synthesized Dy(III) complex using a synthetic method similar to that used for other lanthanide ions. This complex showed a typical Dy(III) ion-based luminescence under irradiation with UV light (315 nm), which confirmed the formation of a Dy(III) complex with ligand L (Figure S1). The molecular structure was revealed by an X-ray diffraction study. The structure of the cationic part of [DyL(NO$_3$)$_2$]PF$_6$·MeCN (**3**) is given in Figure 2. Crystallographic data, accompanied by coordination distances and angles, were summarized in Tables S1 and S2. In **3**, neutral ligand L occupied the equatorial positions around the Dy(III) ion, forming five pentagonal chelating rings. Each bpy moiety maintained its planar arrangement, with the two terminated-pyridine rings in a face-to-face conformation, with shortest contacts of 3.198(5) Å (C1 ... C24). This led to the helical coordination of the entire ligand with a twisting angle of 32.5(1)°, which was defined as the dihedral angle between the two bipyridine moieties. Two nitrate anions were coordinated above and below the Dy(III) ion to complete the deca-coordination. The nitrate–Dy–nitrate arrangement was almost linear, with a N7–Dy–N8 angle of 172.18(8)°. Two nitrate groups were located in a twisting manner around the

N7–Dy–N8 axis with the dihedral angle of 64.30(15)°, which is confirmed by the top view in Figure 2. Anionic nitrate coordinated with a slightly shorter bond distance (Dy–O = 2.451(3)–2.488(3) Å) than the neutral N_6 ligand (Dy–N = 2.495(3)–2.596(3) Å). The estimated coordination distances were slightly shorter than those of the Tb(III) complex (Tb–O = 2.466(3)–2.500(3) Å, Tb–N = 2.505(3)–2.601(3) Å). Although there were slight differences in the coordination distances due to lanthanide contraction, the ligand field geometries of the isostructural complexes **1–3** were very similar, which could lead to an axially stressed ligand field by sandwiching the Ln(III) ion between negatively charged nitrate ligands. This was advantageous for achieving Ising-type magnetic anisotropy in the oblate-type Nd(III), Tb(III), and Dy(III) ions.

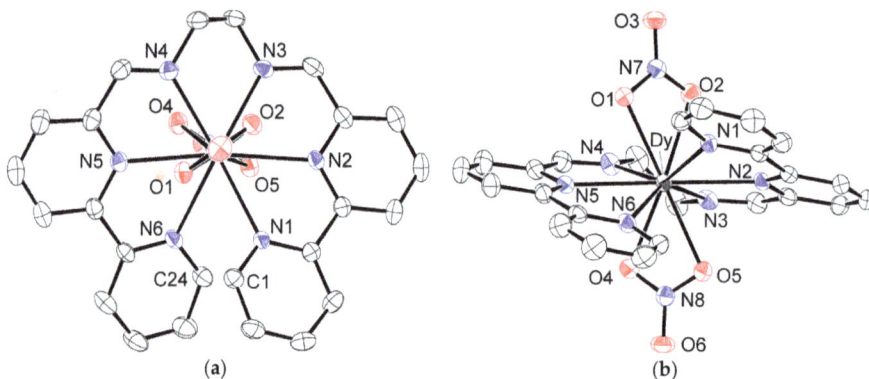

Figure 2. ORTEP drawing of the cationic part of **3** at the 50% probability level. Hydrogen atoms were omitted for clarity: (**a**) top view and (**b**) side view of the molecule.

The complexes were crystalized in a triclinic system with space group *P*-1, and a *Z* value of 2. The nearest molecules were correlated using the inversion center, and the shortest Dy...Dy distances were estimated as 8.7477(12) Å (symmetry code: $-x$, $1-y$, $-z$). The packing diagram in the unit cell is given as Figure S2.

2.2. DC Susceptibility of the Complexes

The magnetic features of the complexes were initially determined from the temperature dependence of $\chi_M T$ products measured under 1000 Oe DC field applied (Figure 3). The expected $\chi_M T$ values of Ln(III) ions in the absence of a crystal field are 1.64 emu·K·mol^{-1} for Nd(III) ($J = 9/2$ and $g = 8/11$), 11.81 emu·K·mol^{-1} for Tb(III) ($J = 6$ and $g = 3/2$), and 14.17 emu·K·mol^{-1} for Dy(III) ($J = 15/2$ and $g = 4/3$). Observed values for **1** (1.64 emu·K·mol^{-1}) and for **2** (12.51 emu·K·mol^{-1}) at room temperature were similar to the expected values; however, the value was slightly smaller for **3** (12.65 emu·K·mol^{-1}) than expected, presumably due to the effect of large magnetic anisotropy. **1** exhibited a Curie-like behavior for the entire temperature range down to 2.0 K, keeping its $\chi_M T$ value constant, however; **2** and **3** showed characteristic dependence of $\chi_M T$ values on temperature when the samples were cooled down to 2.0 K. The $\chi_M T$ value of **3** was almost constant on cooling; below 100 K, it gradually decreased, suggesting the presence of an intrinsic thermal-depopulation process as well as a weak inter-molecular magnetic interaction. The $\chi_M T$ value of **2** linearly decreased on cooling for the entire temperature range, which may be due to the thermal depopulation among several pairs of J_z-sublevels split under an anisotropic crystal field.

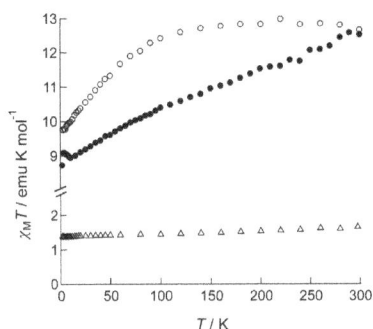

Figure 3. Temperature dependence of the $\chi_M T$ products of **1** (open triangles), **2** (closed circles), and **3** (open circles) measured under an applied DC field of 1000 Oe.

2.3. Dynamic Susceptibility of the Complexes

2.3.1. Nd(III) Complex **1**

The slow magnetic relaxation of the complexes was revealed by measuring the alternating current (AC) magnetic susceptibility under zero field or DC field applied conditions. For Complex **1**, the dynamic behavior of susceptibility was initially measured under a zero DC bias field (Figure S3), which exhibited no out-of-phase signals, χ_M'', due to the quantum-tunneling magnetization (QTM) relaxation process, which was faster than the reversal of the magnetic field. The products of in-phase susceptibility and temperature, $\chi_M'T$, were almost constant at this AC field frequency and exhibited Curie-like behavior. This could have arisen from the relatively large separation between the ground and excited J_z sub-levels, as well as thermal depopulation phenomena not occurring in this temperature range. To suppress fast relaxation via the QTM process, the AC susceptibility was measured under 1000 Oe DC bias field applied conditions (Figure 4) [18]. Under these conditions, **1** exhibited frequency dependent χ_M'' signals at temperatures up to 5.0 K and an AC frequency of up to 10,000 Hz. Figure 4a shows the frequency dependence of both in-phase and out-of-phase susceptibilities as products of temperature. The Cole–Cole plots (Figure 4b) [19] had a semicircular shape. The AC susceptibility data were well fitted with the generalized Debye equations [1,19]:

$$\chi'(\omega) = \chi_S + (\chi_T - \chi_S) \frac{1 + (\omega\tau)^{1-\alpha} \sin(\pi\alpha/2)}{1 + 2(\omega\tau)^{1-\alpha} \sin(\pi\alpha/2) + (\omega\tau)^{2-2\alpha}} \tag{1}$$

$$\chi''(\omega) = (\chi_T - \chi_S) \frac{(\omega\tau)^{1-\alpha} \cos(\pi\alpha/2)}{1 + 2(\omega\tau)^{1-\alpha} \sin(\pi\alpha/2) + (\omega\tau)^{2-2\alpha}}, \tag{2}$$

where χ_T denotes isothermal susceptibility, χ_S denotes adiabatic susceptibility, τ denotes relaxation time at each temperature, and α denotes the distribution of τ (Table S3). The estimated α values were small enough that slow magnetic relaxation occurred via a single process.

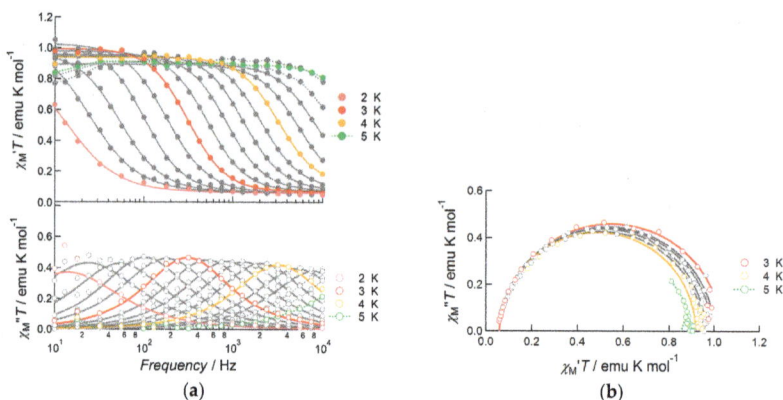

Figure 4. AC susceptibility data of **1** measured under 1000 Oe bias applied field: (**a**) frequency dependence of the products of temperature and in-phase susceptibility (**top**), and temperature and out-of-phase susceptibility (**bottom**), measured at several temperatures from 2.0 to 5.0 K; (**b**) Cole–Cole plots of **1** measured under the same conditions. Solid curves represent theoretical calculations on the basis of generalized Debye equations, of which the estimated parameters are listed in Table S3.

To investigate the slow magnetic relaxation of **1** in detail, the bias field dependence of slow relaxation was revealed in the field range 0–5000 Oe for the temperature range 2.5–4.0 K (Figure 5 and Figure S4). At 2.5 K, no peak was observed when the applied DC field was lower than 200 Oe. By applying a 250 Oe DC field, a small peak appeared at around 70 Hz (τ = 2.3 ms). When the DC field increased from 300 to 2000 Oe, a continuous increase in the amplitude of the out-of-phase signal was observed, keeping the peak frequency at around 70 Hz; above 2500 Oe, the peak frequency was gradually enhanced as the applied field was increased. This field dependence of the peak frequency showed strong temperature dependence when the temperature was increased from 2.5 K to 4.0 K. Above 3.0 K, the flipping rate under 100 Oe was drastically accelerated up to 6000 Hz. Upon increasing the field, the flipping rate decreased, reaching a minimum at around 1000 Oe, and then increased again under a DC field over 2000 Oe. Below 1000 Oe, suppression of the flipping rate under the higher magnetic field indicated that magnetic relaxation occurred via a QTM process. However, strong temperature dependence indicated that the tunneling process was not unique as well as the presence of another possibility. Throughout the whole temperature range outlined above, the relaxation rate was minimal at an applied DC field of 1000 Oe, and so AC data measured under this bias field would be used below in Arrhenius analysis.

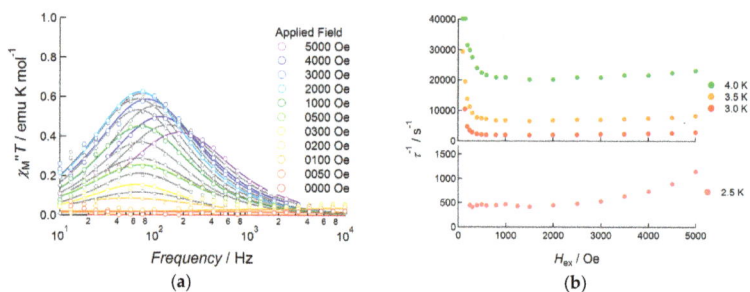

Figure 5. (**a**) Frequency dependence of the out-of-phase component of AC susceptibility of **1** at 2.5 K measured under several DC field applied conditions; (**b**) DC field dependence of the relaxation rate τ^{-1} measured in the temperature range 2.5–4.0 K.

Figure 6 shows the Arrhenius plot for **1** measured under a 1000 Oe DC applied field. A slightly bent plot was obtained over the entire temperature range, which might indicate the presence of several relaxation processes, such as Raman and/or thermally assisted QTM (TA-QTM) processes. Before considering the Raman process, the data were first analyzed using the linear Arrhenius equation for the data above 3.5 K, which gave best fit parameters of $\Delta E/k_B = 36(1)$ K and $\tau_0^{-1} = 4.3(9) \times 10^{-9}$ s. The data were then analyzed using Equation (3), which considers both Raman and TA-QTM processes [20,21]:

$$\tau^{-1} = CT^n + \tau_0^{-1}\exp\left(\frac{-\Delta E}{k_B T}\right).\tag{3}$$

On the right-hand side of the equation, the first term denotes the Raman process, where n takes the values of 5 or 9 for Kramers ions such as Nd(III) and Dy(III). Initially, the data were fitted only considering the Raman process, which gave the value of n as 8.2(1). Then least square fitting based on Equation (3) was carried out for the data, which resulted in better agreement with the observations. The estimated values were $\Delta E/k_B = 34(1)$ K, $\tau_0^{-1} = 1.1(4) \times 10^{-8}$ s, and $C = 2.8(2)$ s^{-1} K^{-5} for $n = 5$, and $\Delta E/k_B = 10(2)$ K, $\tau_0^{-1} = 2.0(16) \times 10^{-4}$ s, and $C = 0.081(2)$ s^{-1} K^{-9} for $n = 9$. The former $\Delta E/k_B$ and τ_0^{-1} values agreed well with the linear analysis results above, and were used to describe the slow magnetic relaxation behavior of **1**.

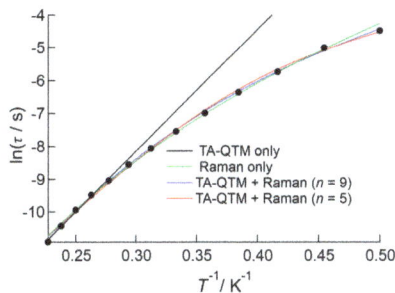

Figure 6. Arrhenius plot of **1** measured under an applied DC field of 1000 Oe. The black line is the result of the fitting to the linear Arrhenius equation, and the green curve is the result of fitting when simply Raman process was considered. The red and black curves represent the results of the fitting using Equation (3), which considers both thermally assisted QTM and Raman processes. The value of n in the Raman term was fixed at 5 for the red curve and 9 for the blue curve.

2.3.2. Tb(III) Complex **2**

Tb(III) complex **2** exhibited no out-of-phase signals under zero field conditions down to 2.0 K (Figure S3). The values of the $\chi_T T$ products were constant and thermal depopulation phenomena were not observed below 7 K. With the application of a DC bias field of 1000 Oe, the temperature- and frequency-dependent out-of-phase component of susceptibility was observed (Figure 7). However, the observed signals were weak and no peaks were observed in this frequency range. This meant that magnetic relaxation occurred at faster frequencies than the flipping rate of the AC field under these conditions.

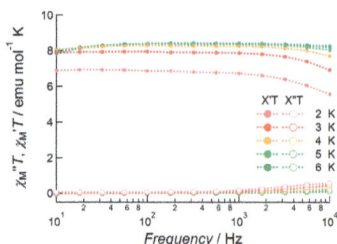

Figure 7. AC susceptibility data of **2** measured under 1000 Oe bias applied field conditions in the temperature range 2.0–6.0 K and frequency range 10–10,000 Hz. Closed circles and open circles denote the value of $\chi_M'T$ and $\chi_M''T$, respectively.

2.3.3. Dy(III) Complex **3**

Under zero field condition, Dy(III) complex **3** exhibited out-of-phase signals around the high frequency limit of field flipping, which indicated the presence of fast relaxation via a tunneling process (Figure S3).

When a 1000 Oe bias field was applied, the fast relaxation process was suppressed and slow magnetic relaxation was observed up to 13.0 K in the frequency range from 10 Hz to 10,000 Hz (Figure 8). To determine the detailed features of slow magnetic relaxation phenomena, both in-phase and out-of-phase susceptibilities were analyzed using generalized Debye Equations (1) and (2) to extract the flipping time τ and distribution of τ, α, as summarized in Table S4. The observed data obeyed Debye equations over the entire temperature range. The α parameter was slightly larger at low temperatures (0.30(1) at 3.0 K and 0.255(6) at 4.0 K), but it was sufficiently small above 5.0 K (0.015(8)–0.196(6) for the temperature range 5.0–12.0 K) indicating a single relaxation process. Hence, **3** was confirmed to be a field-induced SMM. The product of $\chi_T T$ was almost constant in this temperature range and obeyed Curie's law. In contrast, due to the increase in α value on cooling, the $\chi_M''T$ value of each peak decreased when the temperature was reduced from 14.0 to 3.0 K. Relatively large α values below 4.0 K indicated the presence of very weak intermolecular magnetic interactions. We assumed that a through-space dipole–dipole interaction potentially occurred in Dy(III) complex **3**, while in Nd(III) complex **1** this interaction was negligible due to smaller values of J.

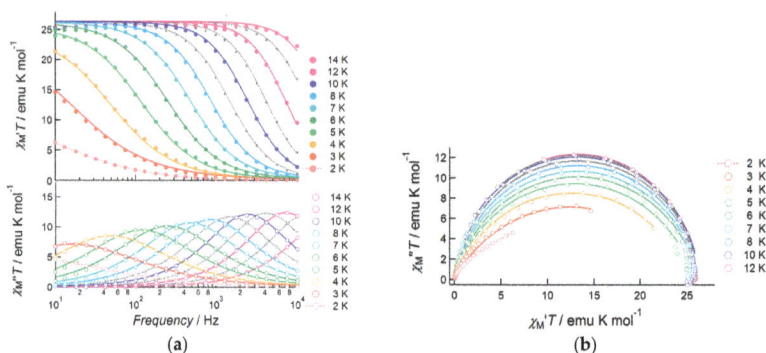

Figure 8. AC susceptibility data for **3**, measured under a 1000 Oe bias applied field: (**a**) frequency dependence of the products of temperature and in-phase susceptibility (**top**), and temperature and out-of-phase susceptibility (**bottom**), measured at several temperatures in the range 2.0–14.0 K; (**b**) Cole–Cole plots of **3** measured under the same conditions. Solid curves represent theoretical calculations on the basis of generalized Debye equations, of which the estimated parameters were listed in Table S4.

To determine the effect of an applied DC field on magnetization flipping, the DC field dependence of AC susceptibility was measured in the temperature range 3.0–7.0 K (Figure 9 and Figure S5). Under zero field conditions at 4.0 K, a shoulder of the out-of-phase signal was observed in the high-frequency region, for which the peak appear at a higher frequency region in the measurement range of the equipment. When a weak DC field was applied, the intensity of the shoulder signal decreased, accompanied by the appearance of a weak peak at around 1000 Hz. Upon increasing the DC field up to 500 Oe, the peak intensity increased with a slowing of the flipping rate. Above 600 Oe, the shoulder signal was disappeared. Above 1000 Oe, each single peak was fitted with Equation (2) using the whole frequency data, whereas below 900 Oe, the selected data around each peak were used for fitting to extract the relaxation rate τ^{-1}. Figure 9b shows the DC field dependence of the relaxation rate. The plot exhibited a clear dependence of τ^{-1} for the field. In the low-field region, the relaxation rate decreased when tunneling relaxation was suppressed by applying a bias field. τ^{-1} maintained its minimum value at around 600–1000 Oe. Then magnetic relaxation was enhanced upon increasing the DC field, which corresponded to the direct process. The data were analyzed using Equation (4) [20,21]:

$$\tau(H, T)^{-1} = AH^4T + \frac{B_1}{1 + B_2H^2} + D, \tag{4}$$

where the first two terms on the right-hand side denote a direct process and the resonance tunneling process, while the third term is a constant at each temperature that includes the terms of thermal relaxation processes, which are independent of DC field strength but dependent on the temperature. The observed data fitted well with Equation (4), giving four parameters: $A = 2.50(5) \times 10^{-12}$ s^{-1} Oe^{-4} K^{-1}, $B_1 = 5.1(5) \times 10^3$ s^{-1}, $B_2 = 6(1) \times 10^{-5}$ Oe^{-2}, and $D = 3.1(5) \times 10^2$ s^{-1} (Table S5).

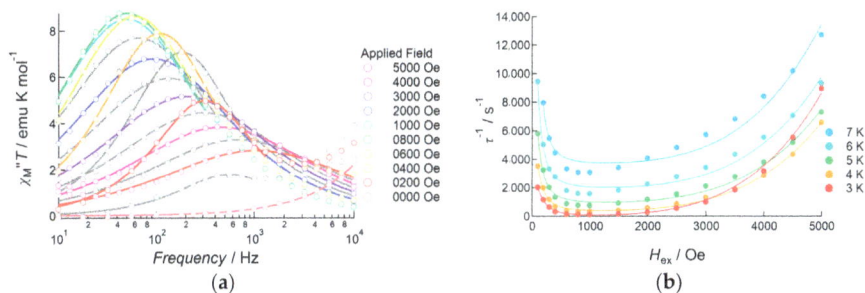

Figure 9. AC susceptibility data for **3**, measured under several DC applied fields: (**a**) frequency dependence of out-of-phase susceptibility at 4.0 K, measured under several DC applied fields ranging from 0 to 5000 Oe. Solid curves represent theoretical calculations on the basis of Equation (2). (**b**) DC field dependence of the relaxation rate measured in the temperature range 3.0–7.0 K. Solid curves represent theoretical calculations on the basis of Equation (4).

Similar plots were obtained in the temperature range 3.0–7.0 K (Figure 9b). At a lower field below 600 Oe, the observed data exhibited strong temperature dependence, but at a higher field, where the direct process was dominant, temperature dependence was smaller. According to Equation (4), in the direct process region, the relaxation rate should increase in a linear manner according to temperature. The least square fitting to Equation (4) was successful below 4.0 K; however, above 5.0 K the observed data did not obey Equation (4), and least square fitting failed (Table S5). This suggested that the magnetic relaxation dynamics of **3** were not explained by this model. To avoid the influence of DC field dependence, Arrhenius analysis was carried out for data measured under the 500 and 1000 Oe DC applied fields, as shown in Figure 10. The relaxation rates measured under the 500 and 1000 Oe DC field conditions were similar in the higher temperature region. However, the difference became larger on cooling. The plots were analyzed above 11.0 K using the linear Arrhenius equation, which resulted in good agreement under both conditions, with values of $\Delta E / k_B = 81(5)$ K and $\tau_0^{-1} = 2.0(9) \times 10^{-8}$ s for 500 Oe applied conditions, and $\Delta E / k_B = 81(5)$ K and $\tau_0^{-1} = 24(10) \times 10^{-9}$ s for 1000 Oe applied conditions. Then, the data were analyzed only considering the Raman process employing C and n as fitting parameters, of which the best fit parameters were summarized in Table S6. These results indicated that the value of n is close to 4 ($H_{ex} = 500$ Oe) or 5 ($H_{ex} = 1000$ Oe); however, the agreement between observations and the theoretical calculations was not sufficient. Finally, the data were analyzed on the basis of Equation (3) both for $n = 5$ and $n = 9$, and applied field of 500 Oe and 1000 Oe conditions. The fitting gave better agreements with the observations; however, the obtained values of the barrier height and relaxations time were inconsistent with the values obtained from the linear Arrhenius analyses: the estimated barrier height was lower, ranging from 7(2) K to 21.8(5) K, and the relaxation time was longer, ranging from $11(15) \times 10^{-5}$ s to $6(3) \times 10^{-4}$ s. Using these values, we can estimate the contribution of the Raman process to the relaxation rate, which reaches more than 50% at 10 K both for $n = 5$ and 9, and increases at higher temperature. This means that the contribution of the Raman process was overestimated in this analysis, and Equation (3) is not appropriate to estimate the barrier height in this case. We will employ the value of $\Delta E / k_B = 81(5)$ K for the discussion given below.

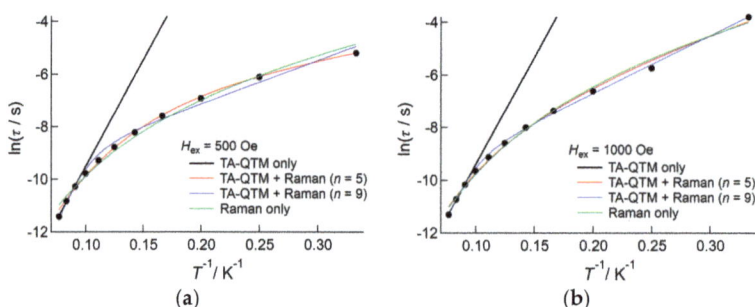

Figure 10. Arrhenius plot of **3**, measured under applied DC fields of (**a**) 500 Oe and (**b**) 1000 Oe. The solid lines and curves represent the results of the fitting based on the linear Arrhenius equation (black) and Raman process (green), and the fitting using Equation (3) with n values of 5 (red) and 9 (blue). The estimated parameters were listed in Table S6.

3. Discussion

Isostructural complexes with different lanthanide(III) centers provided an opportunity to discuss correlations between the coordination structure and magnetic anisotropy. In Complexes **1–3**, oblate-type Nd(III), Tb(III), and Dy(III) ions were located in an axially stressed crystal field, which led to the slow magnetic relaxation phenomena of Nd(III) and Dy(III) complexes. However, any out-of-phase susceptibility was not observed in the Tb(III) complex. There are two possible reasons for these differences. The first is the difference in the quantum number, J. As Tb(III) is a non-Kramers ion with an integer J number, odd numbered J_z sub-levels are not necessarily degenerated, and non-negligible mixing of the sublevels leads to fast tunneling during relaxation. Another possibility is the match/mismatch of crystal field anisotropy and magnetic anisotropy of the series. The $J_z = \pm 6$ sub-levels of Tb(III) were previously reported to have the largest electronic distributions in the equatorial plane, whereas in Nd(III) and Dy(III) ions, each $J_z = \pm 9/2$ and $J_z = \pm 15/2$ sub-level was slightly constricted in the electronic distribution in the equatorial plane. As the N donor atoms surrounded the equatorial plane, the $J_z = \pm J$ sublevels of the Tb(III) ion were more destabilized than those of the Nd(III) and Dy(III) ions, leading to a smaller separation between the ground and excited sub-levels in the Tb(III) complex.

Nd(III) complex **1** showed a field-induced SMM behavior with an effective barrier of $\Delta E / k_B = 34(2)$ K and relaxation rate of $\tau_0^{-1} = 1.1(3) \times 10^{-8}$ s. The smaller J value compared with those of heavy lanthanide ions caused slow magnetic relaxation phenomena to be rarely observed for Nd(III) complexes [13–16,22–24], and a few known examples of Nd(III) complexes show field-induced SMM behaviors. The reported values of $\Delta E/k_B$ varied from 4.09(10) to 40.0(2) K, and the height of the barrier of **1** was similar to that reported. This result also indicated that the synthetic strategy for SMMs, including oblate-type Tb(III) and Dy(III), was valid for the construction of Nd(III)-based SMMs. In the axially ligating nitrate anion, the negative charge is delocalized over the whole moiety. Replacing nitrate anions with carboxylate anions or other monodentate oxygen-donor ligands would enhance the SMM features since the negative charge is more concentrated on donor atom(s) than on nitrate, leading to stronger electronic repulsion.

Dy(III) complex **3** also showed field-induced SMM behavior up to 13.0 K (frequency range: 10–10,000 Hz) with the barrier of $\Delta E/k_B = 81(5)$ K. The estimated barrier of **3** was more than twice as high as that of **1**, mainly due to the larger value of total angular momentum J. The dynamic feature of susceptibility was complicated, showing an unanalyzable dependence on DC bias field strength. The resultant Arrhenius plots showed non-negligible dependence on the DC field, especially at low temperatures, which might partially originate from weak intermolecular interactions. Theoretical calculations, as well as a dilution study of the Dy(III) complex with diamagnetic La(III) or Y(III) complexes, would help to investigate this low-temperature magnetism. At present, La(III) and Y(III) complexes isostructural to the Dy(III) complex are not known, and these studies remain future prospects.

4. Materials and Methods

4.1. General Procedures and Methods

All chemicals and reagents were of reagent grade and used without further purification. All chemical reactions and sample preparations for physical measurements were performed in ambient atmosphere. Variable temperature magnetic susceptibility measurements were performed on MPMS-5S (for DC susceptibility) and PPMS-9 (for AC susceptibility) magnetometers (Quantum Design Japan, Tokyo, Japan). Diamagnetic corrections for each sample were applied using Pascal's constants. The excitation and luminescence spectra of **3** were measured with Fluorolog 3-22 spectrofluorometer (Horiba Jovin Ybon, Kyoto, Japan) at room temperature and 77 K.

4.2. Synthesis of Complexes [LnL(NO₃)₂]PF₆·MeCN

[NdL(NO₃)₂]PF₆·MeCN (**1**) and [TbL(NO₃)₂]PF₆·MeCN (**2**) were prepared according to a previously reported method [12].

[DyL(NO₃)₂]PF₆·MeCN (**3**) was synthesized using a procedure similar to those of **1** and **2**, using Dy(NO₃)₃·6H₂O as the starting material.

4.3. Crystallography

A single crystal of **3** was mounted on a Varimax Saturn area detector (Rigaku Co., Tokyo, Japan) for data collection using confocal monochromated MoKα radiation at low temperature (153 K). Intensity data were corrected for absorption using an empirical method included in the Crystal Clear software [25]. The structures were solved by direct methods with SIR-97 [26], and structure refinement was carried out using the full-matrix least squares method on SHELXL-97 [27]. Non-hydrogen atoms were anisotropically refined and hydrogen atoms were treated using the riding model. Crystallographic data are summarized in Tables S1 and S2. Complete crystal structure results as a CIF file, including bond lengths, angles, and atomic coordinates, are available in the Supplementary Materials. The CCDC number is 1512922 for compound **3**.

Supplementary Materials: The following are available online at www.mdpi.com/2312-7481/2/4/43/s1, Figure S1: excitation and emission spectra of **3**, Figure S2: crystal packing of **3**, Figure S3: frequency dependence of $\chi_M'T$ and $\chi_M''T$ of **1–3**, Figure S4: dc bias field dependence of $\chi_M''T$ of **1**, Figure S5: dc bias field dependence of $\chi_M''T$ of **3**, Table S1: crystallographic data for **3**, Table S2: Coordination distances and angles of **3**, Table S3: best fitted parameters of extended Debye fitting for **1**, Table S4: best fitted parameters of extended Debye fitting for **3**, Table S5: best fitted parameters of **3** using Equation (4), Table S6: best fitted parameters of **3** using Equation (3).

Acknowledgments: This work was supported by a Grant-in Aid for Scientific Research (B) (No. 23350067) and a Grant-in-Aid for Exploratory Research (No. 24655127) from JSPS, Japan (Takashi Kajiwara), as well as by the Exploratory Research Center Project for Private University and a matching fund subsidy from MEXT (2013-2017), Japan (Miki Hasegawa).

Author Contributions: D.I. and M.H. prepared and characterized the complexes. H.W., S.O., and T.K. performed the magnetic measurements and analyzed the data. All the authors reviewed the paper.

Conflicts of Interest: The authors declare no conflict of interest.

Abbreviations

The following abbreviations are used in this manuscript:

SMM	Single Molecule Magnet
QTM	Quantum Tunneling of Magnetization
TM	Thermally Assisted
AC and DC	Alternating and Direct Current

References

1. Gatteschi, D.; Sessoli, R.; Villain, J. *Molecular Nanomagnets*; Oxford University Press: New York, NY, USA, 2006; pp. 47–159.

2. Glaser, T. Rational design of single-molecule magnets: A supramolecular approach. *Chem. Comm.* **2011**, *47*, 116–130. [CrossRef] [PubMed]

3. Layfield, R.A. Organometallic Single-Molecule Magnets. *Organometallics* **2014**, *33*, 1084–1099. [CrossRef]

4. Madhu, N.T.; Tang, J.-K.; Hewitt, I.J.; Clérac, R.; Wernsdorfer, W.; van Slageren, J.; Anson, C.E.; Powell, A.K. What makes a single molecule magnet? *Polyhedron* **2005**, *24*, 2864–2869. [CrossRef]

5. Liddle, S.T.; Van Slageren, J. Improving f-element single molecule magnets. *Chem. Soc. Rev.* **2015**, *24*, 6655–6669. [CrossRef] [PubMed]

6. Feltham, H.L.C.; Brooker, S. Review of purely 4f and mixed-metal nd-4f single-molecule magnets containing only one lanthanide ion. *Coord. Chem. Rev.* **2014**, *276*, 1–33. [CrossRef]

7. Zhang, P.; Guo, Y.-N.; Tang, J. Recent advances in dysprosium-based single molecule magnets: Structural overview and synthetic strategies. *Coord. Chem. Rev.* **2013**, *257*, 1728–1763. [CrossRef]

8. Woodru, D.N.; Winpenny, R.E.P.; Layfield, R.A. Lanthanide Single-Molecule Magnets. *Chem. Rev.* **2013**, *113*, 5110–5148. [CrossRef] [PubMed]

9. Schmitt, D. Angular distribution of 4f electrons in the presence of a crystal field. *J. Phys.* **1986**, *47*, 677–681. [CrossRef]

10. Walter, U. Charge Distributions of Crystal Field States. *Z. Phys. B: Condens. Matter* **1986**, *62*, 299–309. [CrossRef]

11. Rinehart, J.R.; Long, J.R. Exploiting single-ion anisotropy in the design of f-element single-molecule magnets. *Chem. Sci.* **2011**, *2*, 2078–2085. [CrossRef]

12. Hasegawa, M.; Ohtsu, H.; Kodama, D.; Kasai, T.; Sakurai, S.; Ishii, A.; Suzuki, K. Luminescence behaviour in acetonitrile and in the solid state of a series of lanthanide complexes with a single helical ligand. *New J. Chem.* **2014**, *38*, 1225–1234. [CrossRef]

13. Takahara, C.; Then, P.L.; Kataoka, Y.; Nakano, M.; Yamamura, T.; Kajiwara, T. Slow magnetic relaxation of light lanthanidebased linear $LnZn_2$ trinuclear complexes. *Dalton Trans.* **2015**, *44*, 18276–18283. [CrossRef] [PubMed]

14. Hino, S.; Maeda, M.; Kataoka, Y.; Nakano, M.; Yamamura, T.; Kajiwara, T. SMM Behavior Observed in $Ce(III)Zn(II)_2$ Linear Trinuclear Complex. *Chem. Lett.* **2013**, *42*, 1276–1278. [CrossRef]

15. Hino, S.; Maeda, M.; Yamashita, K.; Kataoka, Y.; Nakano, M.; Yamamura, T.; Nojiri, H.; Kofu, M.; Yamamuro, O.; Kajiwara, T. Linear Trinuclear Zn(II)–Ce(III)–Zn(II) Complex which Behaves as Single-molecule Magnet. *Dalton Trans.* **2013**, *42*, 2683–2686. [CrossRef] [PubMed]

16. Le Roy, J.J.; Gorelsky, S.I.; Korobkov, I.; Murugesu, M. Slow Magnetic Relaxation in Uranium(III) and Neodymium(III) Cyclooctatetraenyl Complexes. *Organometallics* **2015**, *34*, 1415–1418. [CrossRef]

17. Singh, S.K.; Gupta, T.; Ungur, L.; Rajaraman, G. Magnetic Relaxation in Single-Electron Single-Ion Cerium(III) Magnets: Insights from Ab Initio Calculations. *Chem. Eur. J.* **2015**, *21*, 13812–13819. [CrossRef] [PubMed]

18. Habib, F.; Long, J.; Lin, P.-H.; Korobkov, I.; Ungur, L.; Wernsdorfer, W.; Chibotaru, L.F.; Murugesu, M. Supramolecular architectures for controlling slow magnetic relaxation in field-induced single-molecule magnets. *Chem. Sci.* **2012**, *3*, 2158–2164. [CrossRef]

19. Cole, K.S.; Cole, R.H. Dispersion and Absorption in Dielectrics I. Alternating Current Characteristics. *J. Chem. Phys.* **1941**, *9*, 341–351. [CrossRef]

20. Abragam, A.; Bleaney, B. *Electron Paramagnetic Resonance of Transition Ions*; Oxford University Press: Oxford, UK, 1970; pp. 60–74 and pp. 555–560.

21. Carlin, R.L. *Magnetochemistry*; Springer: Berlin/Heidelberg, Germany, 1986; pp. 36–51.

22. Vrábel, P.; Orendáč, M.; Orendáčová, A.; Čižmár, E.; Tarasenko, R.; Zvyagin, S.; Wosnitza, J.; Prokleška, J.; Sechovský, V.; Pavlík, V.; et al. Slow spin relaxation induced by magnetic field in [NdCo(bpdo)(H$_2$O)$_4$(CN)$_6$]·3H$_2$O. *J. Phys. Condens. Matter.* **2013**, *25*, 186003. [CrossRef] [PubMed]

23. Zhang, Y.-Z.; Duan, G.-P.; Sato, O.; Gao, S. Structures and magnetism of cyano-bridged grid-like two-dimensional 4f–3d arrays. *J. Mater. Chem.* **2006**, *16*, 2625–2634. [CrossRef]

24. Ma, B.-Q.; Gao, S.; Su, G.; Xu, G.-X. Cyano-Briged 4f-3d Coordination Polymers with a Unique Two-Dimensional Topological Architecture and Unusual Magnetic Behavior. *Angew. Chem. Int. Ed.* **2001**, *40*, 434–437. [CrossRef]

25. *Crystal Clear*, Version 1.3.5; Operating Software for the CCD Detector System; Rigaku and Molecular Structure Corp.: Tokyo, Japan; The Woodlands, TX, USA, 2003.

26. Altomare, A.; Burla, M.C.; Camalli, M.; Cascarano, G.L.; Giacovazzo, C.; Guagliardi, A.; Moliterni, A.G.G.; Polidori, G.; Spagna, R. SIR97: A new tool for crystal structure determination and refinement. *J. Appl. Cryst.* **1999**, *32*, 115–119. [CrossRef]

27. Sheldrick, G.M. *SHELXL-97: Program for the Refinement of Crystal Structures*; University of Göttingen: Göttingen, Germany, 1996.

magnetochemistry

MDPI

Article

Using the Singly Deprotonated Triethanolamine to Prepare Dinuclear Lanthanide(III) Complexes: Synthesis, Structural Characterization and Magnetic Studies [†]

Ioannis Mylonas-Margaritis [1], Julia Mayans [2], Stavroula-Melina Sakellakou [1], Catherine P. Raptopoulou [3], Vassilis Psycharis [3], Albert Escuer [2,*] and Spyros P. Perlepes [1,4,*]

[1] Department of Chemistry, University of Patras, 265 04 Patras, Greece; ioannismylonasmargaritis@gmail.com (I.M.-M.), melinasakell@gmail.com (S.-M.S.)

[2] Departament de Química Inorgànica i Orgànica, Secció Inorgànica, and Institue of Nanoscience and Nanotechnology (IN2UB), Universitat de Barcelona, Av. Diagonal 645, 08028 Barcelona, Spain; julia.mayans@qi.ub.edu

[3] Institute of Nanoscience and Nanotechnology, NCSR "Demokritos", 153 10 Aghia Paraskevi Attikis, Greece; c.raptopoulou@inn.demokritos.gr (C.P.R.); v.psycharis@inn.demokritos.gr (V.P.)

[4] Institute of Chemical Engineering Sciences, Foundation for Research and Technology-Hellas (FORTH/ICE-HT), Platani, P.O. Box 1414, 265 04 Patras, Greece

* Correspondence: albert.escuer@ub.edu (A.E.); perlepes@patreas.upatras.gr (S.P.P.); Tel.: +34-93-403-9141 (A.E.); +30-2610-996-730 (S.P.P.)

[†] This article is dedicated to Dante Gatteschi, a pioneer in the interdisciplinary field of Molecular Magnetism and a great mentor, on the occasion of his 70th birthday.

Academic Editor: Kevin Bernot
Received: 21 December 2016; Accepted: 16 January 2017; Published: 26 January 2017

Abstract: The 1:1 reactions between hydrated lanthanide(III) nitrates and triethanolamine (teaH$_3$) in MeOH, in the absence of external bases, have provided access to the dinuclear complexes [Ln$_2$(NO$_3$)$_4$(teaH$_2$)$_2$] (Ln = Pr, **1**; Ln = Gd, **2**; Ln = Tb, **3**; Ln = Dy, **4**; Ln = Ho, **5**) containing the singly deprotonated form of the ligand. Use of excess of the ligand in the same solvent gives mononuclear complexes containing the neutral ligand and the representative compound [Pr(NO$_3$)(teaH$_3$)$_2$](NO$_3$)$_2$ (**6**) was characterized. The structures of the isomorphous complexes **1**·2MeOH, **2**·2MeOH and **4**·2MeOH were solved by single-crystal X-ray crystallography; the other two dinuclear complexes are proposed to be isostructural with **1**, **2** and **4** based on elemental analyses, IR spectra and powder XRD patterns. The IR spectra of **1–6** are discussed in terms of structural features of the complexes. The two Ln^{III} atoms in centrosymmetric **1**·2MeOH, **2**·2MeOH and **4**·2MeOH are doubly bridged by the deprotonated oxygen atoms of the two $\eta^1:\eta^1:\eta^1:\eta^2:\mu_2$ teaH$_2^-$ ligands. The teaH$_2^-$ nitrogen atom and six terminal oxygen atoms (two from the neutral hydroxyl groups of teaH$_2^-$ and four from two slightly anisobidentate chelating nitrato groups) complete 9-coordination at each 4f-metal center. The coordination geometries of the metal ions are spherical-relaxed capped cubic (**1**·2MeOH), Johnson tricapped trigonal prismatic (**2**·2MeOH) and spherical capped square antiprismatic (**4**·2MeOH). O–H···O H bonds create chains parallel to the *a* axis. The cation of **6** has crystallographic two fold symmetry and the rotation axis passes through the Pr^{III} atom, the nitrogen atom of the coordinated nitrato group and the non-coordinated oxygen atom of the nitrato ligand. The metal ion is bound to the two $\eta^1:\eta^1:\eta^1:\eta^1$ teaH$_3$ ligands and to one bidentate chelating nitrato group. The 10-coordinate Pr^{III} atom has a sphenocoronal coordination geometry. Several H bonds are responsible for the formation of a 3D architecture in the crystal structure of **6**. Complexes **1–6** are new members of a small family of homometallic Ln^{III} complexes containing various forms of triethanolamine as ligands. Dc magnetic susceptibility studies in the 2–300 K range reveal the presence of a weak to moderate intramolecular antiferromagnetic exchange interaction ($J = -0.30(2)$ cm^{-1} based on the spin Hamiltonian $\hat{H} = -J(\hat{S}_{Gd1} \cdot \hat{S}_{Gd1'})$) for **2** and probably weak antiferromagnetic exchange interactions within the molecules of **3–5**. The antiferromagnetic $Gd^{III} \cdots Gd^{III}$ interaction in **2** is discussed in terms of known magnetostructural correlations for complexes possessing the $\{Gd_2(\mu_2\text{-}OR)_2\}^{4+}$ core. Ac magnetic susceptibility measurements in zero dc field for **3–5** do not show frequency dependent out-of-phase signals; this experimental fact is discussed and rationalized for complex **4** in terms of the magnetic anisotropy axis for each Dy^{III} center and the oblate electron density of the metal ion.

Keywords: dinuclear lanthanide(III) complexes; ground state magnetic axes of dysprosium(III) ions in a complex; magnetic properties; triethanolamine-lanthanide(III) complexes

1. Introduction

Electrons residing in 4f orbitals give trivalent lanthanides (Ln^{III}) interesting (and sometimes difficult to understand in detail) optical and magnetic properties that are currently exploited in modern technology. As far as magnetism is concerned, Ln^{III} ions have been the key components of hard permanent magnets, such as the SmCo and FeNdB magnets, as well as the platform to investigate interesting physical phenomena, e.g., the spin-ice behavior in $Dy_2Ti_2O_7$ [1]. Ln^{III} complexes have also attracted intense interest since the first days of Molecular Magnetism, because of their high magnetic moments and large anisotropies [2]. However, it was after the discovery that magnetic anisotropy could lead to magnetic hysteresis at the molecular level in a class of compounds known as Single-Molecule Magnets (SMMs) [3,4] that an explosion in their study has occurred. In 2003, Ishikawa's group reported SMM behavior in the double-decker phthalocyanine complexes $(Bu^n_4N)[Ln(pc)_2]$ (Ln = Tb, Dy) [5], which gave birth of the Ln^{III} SMM era [6,7]. The magnetization dynamics of dinuclear and polynuclear Ln^{III} SMMs mainly originate from single-ion behavior because it is difficult to create effective pathways for magnetic interactions between metal centers due to the nature of inner 4f electrons; in general, both the exchange and dipolar interactions between Ln^{III} ions are weak [8]. This situation is different from the magnetic relaxation of transition-metal-based SMMs, in which both the spin and exchange interactions contribute simultaneously to the magnetization dynamics [3]. As an alternative candidate, Ln^{III} centers that are usually weakly coupled in dinuclear and polynuclear complexes have shown advantages with respect to SMM studies because large magnetization reversal energy barriers can be achieved via single-ion anisotropy originating from strong spin-orbit coupling and crystal field effects [6,8,9].

Dinuclear complexes represent the simplest molecular entities which allow the study of magnetic interactions between Ln^{III} spin carriers [10]. By studying such systems, researchers could expect to understand the nature and strength of $Ln^{III} \cdots Ln^{III}$ exchange interactions, as well as possible alignment of spin vectors and anisotropy axes. These parameters can be affected by the molecular symmetry, the coordination geometry and/or the bridging ligands, which may act as superexchange pathways. More importantly, dinuclear Ln^{III} SMMs are very important model systems to answer basic questions regarding single-ion relaxation vs. slow magnetic relaxation arising from the molecule as an entity. Thus, Ln^{III}_2 complexes are highly desirable [10–20]. Another interesting area to which dinuclear lanthanide(III) complexes (and also mononuclear ones) are relevant is quantum computation; Ln^{III} ions are promising candidates for encoding quantum information [21,22]. For the realization of a quantum gate, asymmetric dinuclear molecules composed of two weakly coupled Ln^{III} qubits are promising [21]. However, the synthesis of asymmetric molecular dimers is not straightforward, as nature tends to make them symmetric. From the synthetic inorganic chemistry viewpoint, the most logical simple route for the isolation of dinuclear 4f-metal ion complexes is the simultaneous employment of bidentate bridging *anionic* groups (e.g., $\eta^1{:}\eta^1{:}\mu_2$ and/or $\eta^1{:}\eta^2{:}\mu_2$ carboxylate groups) and chelating (most often bidentate or tridentate, e.g. bpy, phen, terpy, etc.) *neutral* capping organic ligands, which terminate oligomerization or polymerization by blocking two or three coordination sites per Ln^{III} center [23]. Another method is the simultaneous employment of capping bidentate nitrato groups (nitrato ligands have little tendency for bridging in Ln^{III} chemistry) and *neutral or anionic* organic ligands that can, in principle, bridge only two metal centers; in addition to the bridging functionality, the ligands should preferably possess "chelating" parts to satisfy the demand for high coordination numbers at the Ln^{III} centers [24–26]. Thus, the choice of the primary organic ligands is crucial for the synthesis of Ln$^{III}_2$ complexes.

With all the above in mind and given the recently initiated interest of our groups in Ln$^{III}_2$ complexes with interesting magnetic and/or luminescence properties [23,25–27], we report here the synthesis, structures and magnetic properties of new such complexes bearing the monoanion of tris(2-hydroxyethyl)amine (the empirical name is triethanolamine, abbreviated hereafter as teaH$_3$, Scheme 1). This ligand is very popular in 3d- and mixed 3d/4f-metal cluster chemistry (representative 3d/4f-metal compounds based on anionic forms of teaH$_3$ are described in refs [28–31]), but with limited used in homometallic Ln^{III} and Ln^{II} (Ln^{II} = Eu^{II}, Yb^{II}) chemistry. Ln^{III} and Ln^{II} complexes containing only the neutral ligand are always mononuclear [32–36], whereas the doubly deprotonated form (teaH$^-$) of the ligand, either alone or with the simultaneous presence of other bridging inorganic or organic ligands, have led to high nuclearity Ln^{III} clusters [37–39]. The singly deprotonated (teaH$_2^-$) ligand has been employed only with combination with other bridging groups and such ligand "blends" give polynuclear Ln^{III} complexes [38–40], although the teaH$_2^-$ group usually (but not always) bridges two metal centers.

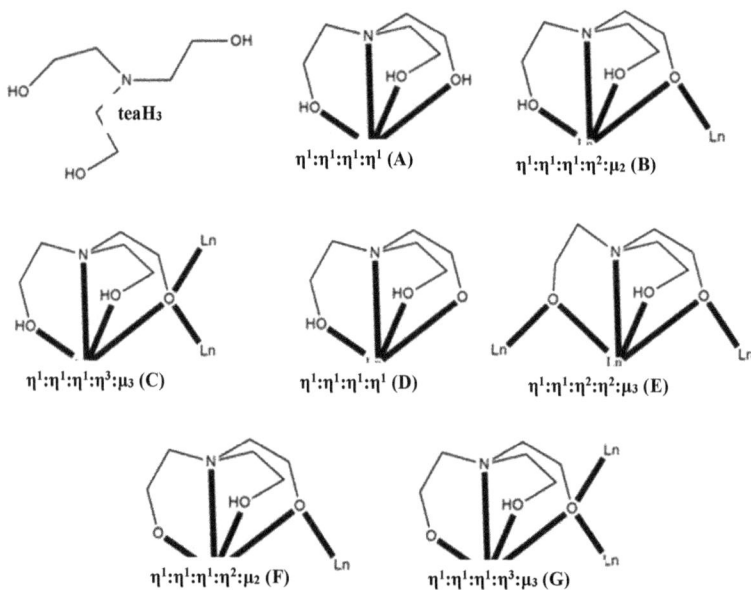

Scheme 1. The structural formula of teaH$_3$ (up, left) and the to date crystallographically established coordination modes of the: neutral (**A**); and monoanionic (**B–D**) teaH$_2^-$; and dianionic (**E–G**) teaH^{2-} ligands in *homometallic* Ln^{II} (Ln = Eu, Yb) and Ln^{III} chemistry.

2. Results and Discussion

2.1. Synthetic Comments and IR Discussion

Since we were interested in isolating dinuclear lanthanide(III) complexes with the *singly deprotonated* form of triethanolamine, i.e., teaH$_2^-$, as ligand, we avoided the addition of external bases (Et$_3$N, LiOH, Me$_4$NOH, etc.) in the systems. Previous reports have shown that use of Et$_3$N in the Ln^{III}/teaH$_3$ reaction mixtures in MeOH and MeOH/CH$_2$Cl$_2$ lead to clusters [Ln$_6$(NO$_3$)$_6$(teaH)$_6$] [37]

and $[Ln_8(OH)_6(teaH)_6(teaH_2)_2(teaH_3)_2](O_3SCF_3)_4$ [39], respectively, containing exclusively or partially the dianionic teaH^{2-} ligand. We hoped that the formation of Ln^{III}–O(teaH$_3$) bonds in solution would lead to moderate polarization of the –CH$_2$–CH$_2$–O–H groups of teaH$_3$ and subsequent single deprotonation of the ligand; this has, indeed, turned out to be the case. Thus, the 1:1 Ln(NO$_3$)$_3$·xH$_2$O/teaH$_3$ reaction mixtures in MeOH gave solutions from which were subsequently isolated crystals of the dinuclear complexes $[Ln_2(NO_3)_4(teaH_2)_2]$ (Ln = Pr, **1**; Ln = Gd, **2**; Ln = Tb, **3**; Ln = Dy, **4**; Ln = Ho, **5**) in yields of ca. 50%. The crystals of **1**, **2** and **4** (in the form of their bis-methanol solvates) were of X-ray quality and their structures were solved by single crystal X-ray crystallography. Complexes **3** and **5** (isolated as polycrystalline powders) are proposed to be isostructural with **1**, **2** and **4** based on elemental analyses and IR spectra (vide supra), as well as on powder XRD patterns (vide infra, Figure S1) for **2**, **3** and **4**. Assuming that the dinuclear complexes are the only products from the general Ln(NO$_3$)$_3$·xH$_2$O/teaH$_3$ (1:1) reaction system in MeOH, their formation can by summarized by Equation (1), where x is 6 or 5:

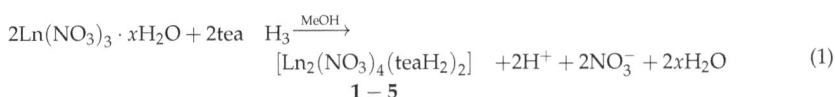

$$2Ln(NO_3)_3 \cdot xH_2O + 2tea \quad H_3 \xrightarrow{\text{MeOH}}$$
$$\underset{\mathbf{1-5}}{[Ln_2(NO_3)_4(teaH_2)_2]} \quad +2H^+ + 2NO_3^- + 2xH_2O \qquad (1)$$

Analogous reactions in MeCN gave amorphous powders which we could not characterize.

In a next step, we examined the influence of the teaH$_3$: Ln^{III} reaction ratio on the product identity. Using an excess of the ligand, i.e., teaH$_3$:Ln(NO$_3$)$_3$·xH$_2$O molar ratios of 2.5:1, we anticipated the formation of complexes of the general formula $[Ln_2(NO_3)_2(teaH_2)_4]$. Somewhat to our disappointment, the products of such reactions in MeOH with the lighter lanthanides are the mononuclear complexes $[Ln(NO_3)(teaH_3)_2](NO_3)_2$, suggesting that the excess of teaH$_3$ does not favor the efficient polarization (due to coordination) of the –CH$_2$–CH$_2$–O–H groups of the ligand that would lead to its single deprotonation. To fully characterize the mononuclear complexes, we have crystallized the representative Pr^{III} complex and solved its structure (vide infra); its formation is summarized by Equation (2):

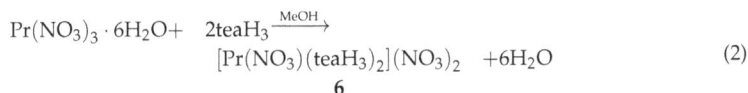

$$Pr(NO_3)_3 \cdot 6H_2O + \quad 2teaH_3 \xrightarrow{\text{MeOH}}$$
$$\underset{\mathbf{6}}{[Pr(NO_3)(teaH_3)_2](NO_3)_2} \quad +6H_2O \qquad (2)$$

IR evidence shows that the mononuclear complexes with the heavier lanthanide(III) ions, e.g., Dy^{III}, Ho^{III} and Er^{III}, are probably best formulated as eight-coordinate $[Ln(teaH_3)_2](NO_3)_3$ as a consequence of the lanthanide contraction. Since we were not particularly interested in mononuclear Ln^{III}/teaH$_3$ complexes, we did not pursue this chemistry further.

In the IR spectra of well-dried samples of complexes **1–5**, the strong intensity broad band at ~3350 cm^{-1} is assigned to the ν(OH) vibration of the coordinated –OH groups of teaH$_2^-$; this mode in the spectrum of **6** appears as two bands at ~3400 and 3230 cm^{-1} [41]. The broadness of the bands is indicative of the participation of the hydroxyl groups in H-bonding. The bands at ~1650 and 1473–1458 cm^{-1} are tentatively assigned to the δ(OH) and ν(C–C) modes, respectively. The KBr IR spectra of **1–5** exhibit a strong band at 1384 cm^{-1}, characteristic of the $\nu_3(E')$ [ν_d(N–O)] mode of the planar D_{3h} ionic nitrate [42,43]; such a nitrate is absent in the complexes. The appearance of this band suggests that the nitrato ligands are replaced by bromides that are present in excess in the spectroscopic KBr matrix, thus producing ionic nitrates (KNO$_3$); this replacement is facilitated by the pressure used for the preparation of the pellet [42,43]. In accordance with this conclusion, the band at 1384 cm^{-1} is absent from the mull spectra of these complexes. The mull IR bands at ~1480 and ~1290 cm^{-1} in **1–5** are assigned [25] to the $\nu_1(A_1)$[ν(N=O)] and $\nu_5(B$textsubscript2$)$[ν_{as}(NO$_2$)] vibrational modes, respectively, of the bidentate chelating (C_{2v}) nitrato group. The separation of these two highest-wavenumber stretching bands (~190 cm^{-1}) is large and typical of bidentate nitrates [25,42,43]. The KBr IR spectrum of **6** (which contains both chelating and ionic nitrates) exhibits only the ionic nitrate (D_{3h}) band at 1383 cm^{-1} due to the bidentate nitrato group → ionic nitrate transformation mentioned above; however, the mull spectra (nujol and hexachlorobutadiene) show the typical bands of both the bidentate nitrato group (at 1490 and 1285 cm^{-1}) and ionic nitrate (1383 cm^{-1}), as expected. The IR spectra (KBr) of the representative dinuclear complex **3** and the free teaH$_3$ ligand (liquid between CsI plates) are presented in Figures S2 and S3, respectively.

2.2. Description of Structures

The structures of **1**·2MeOH, **2**·2MeOH, **4**·2MeOH and **6** have been solved by single-crystal X-ray crystallography (Table 1). Complexes **1**·2MeOH, **2**·2MeOH and **4**·2MeOH are isomorphous and thus only the structure of **4**·2MeOH will be described in detail. Complexes **3** and **5** are proposed to have similar structures with those of **1**, **2** and **4** based on elemental analyses, IR spectra and powder X-ray patterns (Figure 1 and Figure S1). The measured pXRD patterns of **2**, **3** and **4** are very similar indicating that the complexes are isostructural. The differences in intensity may be due to the preferred orientation of the crystalline powder samples [44,45]. The experimental patterns agree satisfactorily with those calculated from the single-crystal X-ray diffraction data. However, they are not identical; a possible reason for this is the fact that the measured patterns correspond to unsolvated complexes, i.e., **2**, **3** and **4** (this is also confirmed by elemental analyses), while the simulated ones refer to the bis(methanol) solvates. Structural plots of the compounds **4**·2MeOH and **6**, as well as supramolecular features of complex **4**·2MeOH, are presented in Figures 2–4, while numerical data are listed in Tables 2 and 3 and Tables S1–S6. Structural plots of **1**·2MeOH and **2**·2MeOH are shown in Figures S4 and S5, respectively.

Table 1. Crystallographic data for complexes **1**·2MeOH, **2**·2MeOH, **4**·2MeOH and **6**.

Parameter	1·2MeOH	2·2MeOH	4·2MeOH	6
Formula	$C_{14}H_{36}N_6Pr_2O_{20}$	$C_{14}H_{36}N_6Gd_2O_{20}$	$C_{14}H_{36}N_6Dy_2O_{20}$	$C_{12}H_{30}N_5PrO_{15}$
Formula weight	890.31	922.99	933.49	652.32
Crystal system	triclinic	triclinic	triclinic	monoclinic
Space group	$P\bar{1}$	$P\bar{1}$	$P\bar{1}$	$C2/c$
Radiation	Mo Kα	Mo Kα	Mo Kα	Mo Kα
T/K	160	230	160	160
$a/\text{Å}$	8.3271(2)	8.3299(4)	8.2897(4)	11.5500(2)
$b/\text{Å}$	8.6978(2)	8.6424(4)	8.6055(4)	14.8428(3)
$c/\text{Å}$	10.3787(3)	10.3733(5)	10.2855(5)	13.8550(3)
$\alpha/°$	86.893(1)	86.964(2)	86.892(1)	90
$\beta/°$	80.373(1)	79.310(1)	79.141(1)	107.393(1)
$\gamma/°$	84.632(1)	84.494(2)	84.532(1)	90
$V/\text{Å}^3$	737.30(3)	729.96(6)	716.83(6)	2266.62(8)
Z	1	1	1	4
$D_{calc}/\text{g·cm}^{-3}$	2.005	2.100	2.162	1.832
μ/mm^{-1}	3.36	4.60	5.27	2.23
Reflns measured	15028	13695	9845	21716
Reflns unique (R_{int})	3210(0.018)	3184(0.024)	3119(0.026)	2463(0.021)
Reflns with I > 2σ(I)	3143	3099	3042	2424
GOF on F^2	1.12	1.07	1.07	1.08
$R_1{}^a$ [I > 2σ(I)]	0.0129	0.0129	0.0142	0.0134
$wR_2{}^b$ (all data)	0.0305	0.0292	0.0332	0.0347

$^a R_1 = \Sigma(|F_o| - |F_c|)/\Sigma(|F_o|); ^b wR_2 = \left\{ \Sigma\left[w(F_o^2 - F_c^2)^2 \right] / \Sigma\left[w(F_o^2)^2 \right] \right\}^{\frac{1}{2}}.$

Complex **4**·2MeOH crystallizes in the triclinic space group $P\bar{1}$. Its structure consists of dinuclear [Dy$_2$(NO$_3$)$_4$(teaH$_2$)$_2$] and solvate MeOH molecules in an 1:2 ratio; the latter will not be further discussed. The asymmetric unit contains half of the dinuclear complex molecule and one MeOH lattice solvent. The dinuclear molecule possesses an inversion center at the mid-point of the Dy1···Dy1′ distance (3.669(1) Å). The two Dy^{III} atoms are doubly bridged by the deprotonated oxygen atoms

(O1, and O1′) of the two $\eta^1{:}\eta^1{:}\eta^1{:}\eta^2{:}\mu_2$ teaH$_2^-$ ligands (Scheme 1B). The teaH$_2^-$ nitrogen atom (N1) and six terminal oxygen atoms (two from the neutral hydroxyl groups of the organic ligand and four from two slightly anisobidentate chelating nitrato groups) complete 9-coordination at each metal center. The Dy–O/N bond lengths are typical of nine-coordinate Dy^{III} atoms [27,46]. The Dy1–O1,O1′ bond lengths (2.265(1), and 2.236(1) Å) involving the deprotonated oxygen atoms are shorter than the Dy1–O2,O3 bond distances (2.363(1), and 2.466(2) Å) involving the neutral hydroxyl atoms of teaH$_2^-$, as expected. Due to the bidentate coordination of the nitrato lligands, the N2–O6 (1.233(2) Å) and N3–O9 (1.216(2) Å) bond distances involving the "free", i.e., uncoordinated, oxygen atoms are shorter than the N2–O4, O5 (1.261(2) Å) and N3–O7,O8 (1.263(2), and 1.280(2) Å) distances involving the coordinated oxygen atoms.

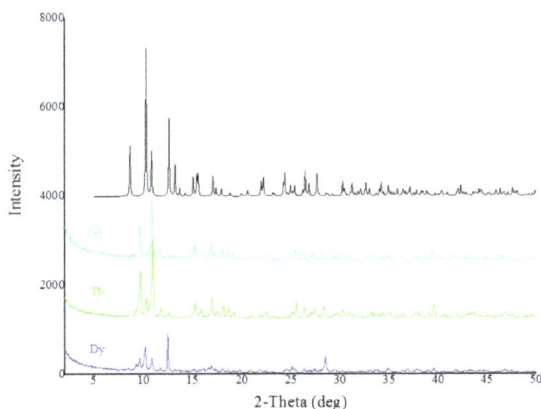

Figure 1. Experimental X-ray diffraction patterns of complexes **2** (cyan), **3** (green) and **4** (blue), and the theoretical pattern for complex **2** (black).

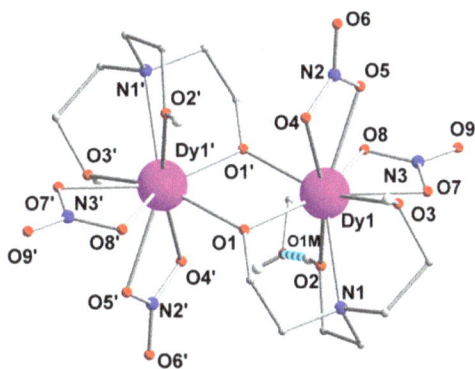

Figure 2. Partially labeled plot of the molecule [Dy$_2$(NO$_3$)$_4$(teaH$_2$)$_2$] that is present in the structure of **4**·2MeOH. One lattice MeOH molecule is also shown. The thick dotted cyan line represents the O2–H(O2)···O1M H bond. Symmetry operation used to generate equivalent atoms: (′) $-x+1, -y, -z+1$. Only the H atoms at O2 and O3 (and their centrosymmetric equivalents) are shown.

Table 2. Selected interatomic distances (Å) and bond angles (°) for the representative complex [Dy$_2$(NO$_3$)$_4$(teaH$_2$)$_2$]·2MeOH (4·2MeOH).

Interatomic Distances (Å) [a]			
Dy1···Dy1′	3.669(1)	Dy1–O8	2.527(2)
Dy1–O1	2.265(1)	Dy1–N1	2.653(2)
Dy1···O1′	2.236(1)	N2–O4	1.261(2)
Dy1–O2	2.363(1)	N2–O5	1.261(2)
Dy1–O3	2.466(2)	N2–O6	1.233(2)
Dy1–O4	2.564(2)	N3–O7	1.280(2)
Dy1–O5	2.492(2)	N3–O8	1.263(2)
Dy1–O7	2.463(2)	N3–O9	1.216(2)

Bond Angles (Å) [a]			
O1–Dy1–O1′	70.8(1)	O5–Dy1–O1	122.3(1)
O1–Dy1–O2	95.5(1)	O5–Dy1–O7	73.7(1)
O1–Dy1–O7	157.3(1)	O7–Dy1–O8	51.2(1)
O1′–Dy1–O3	149.6(1)	O8–Dy1–O5	67.7(1)
O2–Dy1–O5	136.2(1)	N1–Dy1–O2	65.8(1)
O2–Dy1–O7	78.4(1)	N1–Dy1–O5	144.5(1)
O3–Dy1–O5	77.9(1)	Dy1–O1–Dy1′	109.2(1)
O3–Dy1–O8	114.8(1)	O4–N2–O5	117.4(2)
O4–Dy1–O5	50.4(1)	O4–N2–O6	120.9(2)
O4–Dy1–O7	113.2(1)	O5–N2–O6	121.7(2)

[a] Symmetry operation used to generate equivalent atoms: (′) −x + 1, −y, −z + 1.

Table 3. Selected bond lengths (Å) and angles (°) for complex [Pr(NO$_3$)(teaH$_3$)$_2$](NO$_3$)$_2$.

Bond Lengths (Å)			
Pr1–O1	2.514(1)	N2–O5	1.224(2)
Pr1–O2	2.502(1)	N1–C1A	1.487(4)
Pr1–O3	2.521(1)	C1A–C2A	1.518(4)
Pr1–O4	2.584(1)	C2A–O1	1.429(2)
Pr1–N1	2.743(1)	C3–O2	1.434(2)
N2–O4	1.271(1)	C5–O3	1.430(2)

Bond Angles (°) [a]			
O1–Pr1–O2	115.6(1)	O4–Pr1–O4′	49.6(1)
O1–Pr1–O3	77.1(1)	O4–Pr1–N1	82.5(1)
O1–Pr1–O4	69.6(1)	O4′–Pr1–N1	122.5(1)
O1–Pr1–N1	63.4(1)	O4–N2–O4′	116.9(2)
O2–Pr1–O2′	172.0(5)	O4–N2–O5	121.6(1)
O3–Pr1–O3′	70.2(1)	C1A–C2A–O1	107.4(2)
N1–Pr1–Nq′	154.1(1)	C6–C5–O3	108.1(1)

[a] Symmetry operation used to generate equivalent atoms: (′) −x + 1, y, −z + 1/2.

To estimate the closer coordination polyhedron defined by the donor atoms around Dy1 (and its centrosymmetric equivalent), a comparison of the experimental structural data with the theoretical data for the most common polyhedral structures with nine vertices was performed by means of the program SHAPE [47]. The so-called Continuous Shape Measures (CShM) approach essentially allows one to numerically evaluate by how much a particular structure deviates from an ideal shape. As there are no Platonic, Archimedean or Catalan polyhedra with nine vertices, and as no prisms or antiprisms can be constructed with an odd number of vertices, the main semiregular three-dimensional figures that may be considered are those listed (except, of course, the enneagon) in Table S1. The best fit (Table S1) was obtained for the spherical capped square antiprism (Figure S6), with the nitrato oxygen atom O4 being the capping atom.

The lattice structure of the complex is built through H bonds (Table S5) involving the MeOH oxygen atom (O1M) and the neutral hydroxyl oxygen atoms (O2, and O3) as donors, and the O1M, O3 and nitrato oxygen atoms O4 and O6 as acceptors; thus, O1M and O3 act as both donors and acceptors of H bonds. A weak, non-classical H bond is also formed with the teaH$_2$$^-$ carbon atom C2 (this is the atom that is connected to N1) as donor and the nitrato oxygen atom O8 as acceptor. The O–H···O H bonds create chains parallel to the *a* axis (Figure 3).

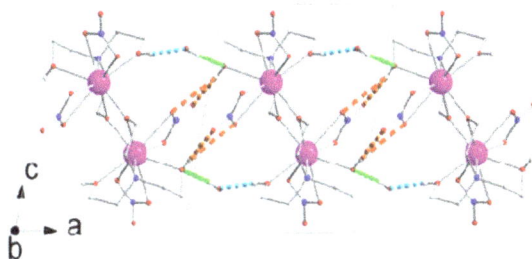

Figure 3. Chains of molecules formed parallel to the *a* axis in the crystal structure of **4**·2MeOH. The thick dashed lines represent the O2–H(O2)···O1M (cyan), O1M–H(O1M)···O3 (light green), O3–H(O3)···O6 and O3–H(O3)···O4 (orange) H bonds.

The Ln–O/N bond lengths and the Ln···Ln distances in the isomorphous complexes **1**·2MeOH, **2**·2MeOH and **4**·2MeOH follow the order Dy < Gd < Pr (for example the Dy···Dy, Gd···Gd and Pr···Pr distances are 3.669(1), 3.719(1) and 3.810(1) Å, respectively), a typical consequence of the lanthanide(III) contraction [25]. The coordination polyhedra of the Gd^{III} (**2**·2MeOH) and Pr^{III} (**1**·2MeOH) are closest to Johnson tricapped trigonal prism and to spherical-relaxed capped cube, respectively (Figures S7 and S8, respectively; Tables S2 and S3, respectively). At first glance, the fact that the Ln^{III} centers in the isomorphous complexes **1**·2MeOH, **2**·2MeOH and **4**·2MeOH have different coordination geometries (and not consistent CShM values) seems strange. We attribute this to two factors: (i) For a given ligand set of nine donor atoms, the tricapped trigonal prism, capped square antiprism and capped cube polyhedra have comparable energies [48,49] and there exist minimal distortion interconversion paths between them; thus many structures are intermediate between two ideal shapes [48], and (ii) the Ln^{III}-donor atom distances are slightly different due to lanthanide(III) contraction and this can affect the shape. For example, the CShM values of the Dy^{III} center in **4**·2MeOH for the spherical capped square antiprism (3.263), spherical-relaxed capped cube (4.016) and tricapped trigonal prism (4.165) are all low and similar (Table S1), and its polyhedron could be equally well described as spherical-relaxed capped cube (a polyhedron that gives the lowest CShM value for the Pr^{III} center in **1**·2MeOH, Table S3).

Complex **6** crystallizes in the monoclinic space group *C*2/*c*. Its structure consists of mononuclear [Pr(NO$_3$)(teaH$_3$)$_2$]$^{2+}$ cations and NO$_3^-$ counterions in a 1:2 ratio; the latter will not be further discussed. The asymmetric unit contains half of the cation and one nitrate anion. The cation has crystallographic twofold symmetry, with Pr1 and the nitrate atoms N2 and O5 occupying the rotation axis. Pr1 is coordinated to two neutral η^1:η^1:η^1:η^1 teaH$_3$ ligands (A in Scheme 1) and to one bidentate chelating nitrato group, and its coordination number is thus 10. The Pr–O/N bond lengths are slightly shorter than the corresponding La–O/N ones in the isomorphous compound [La(NO$_3$)(teaH$_3$)$_2$](NO$_3$)$_2$ [35]. Again, the N2–O5 (1.224(2) Å) bond distance involving the uncoordinated nitrato oxygen atom is shorter than the N2–O4, O4′ distances (1.271(1) Å) involving the bound nitrato oxygen atoms.

Of the accessible 10-coordinate polyhedra for metal ions, the sphenocorona (tetradecahedron) (Figure 4b) is the most appropriate for the description of the 10 donor atoms in **6** according to the program SHAPE [47] (Table S4).

Intercationic interactions through the C3–HB(3)···O4 (and its symmetry equivalent) H bond result in the formation of chains parallel to the *c* axis (Figure 5a). Attached to these chains are lattice NO$_3^-$ ions through the O1–H(O1)···O6 and O2–H(O2)···O8 H bonds (Figure 5a). These lattice NO$_3^-$ ions interact further through the O3–H(O3)···O8 and C6–HA(C6)···O8–H bonds with neighboring chains forming a 3D architecture(Figure 5b). The dimensions of the H bonds are listed in Table S6.

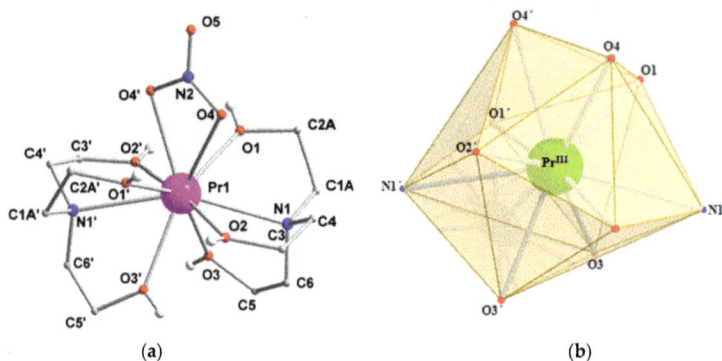

(a)   (b)

Figure 4. (a) Labeled plot of the cation [Pr(NO$_3$)(teaH$_3$)$_2$]$^{2+}$ that is present in the structure of **6**. Since the methylene groups of teaH$_3$ defined by the C1 and C2 atoms present disorder at two sites, only the carbon atoms of the sites with the 0.8 occupancy have been drawn. Symmetry operation used to generate equivalent atoms: (′) −x + 1, y, −z + 1/2. Only the H atoms of the hydroxyl groups are shown. (b) The sphenocoronal coordination geometry of Pr1 in the structure of **6**. The plotted polyhedron is the ideal, best-fit polyhedron using the program SHAPE [47]. Primed and unprimed donor atoms are related by the symmetry operation −x + 1, y, −z + 1/2.

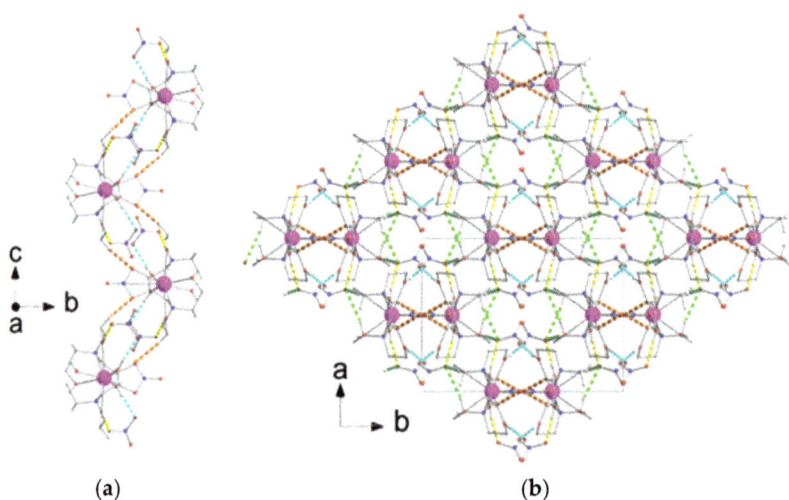

(a)   (b)

Figure 5. (a) Chains of [Pr(NO$_3$)(teaH$_3$)$_2$]$^{2+}$ cations and NO$_3^-$ counterions parallel to the c axis in the crystal structure of **6**. (b) 3D arrangement of chains in the crystal structure of **6**. The thick dashed lines represent the C3−HB(C3)···O4 (orange), O1−H(O1)···O6 (cyan), O2−H(O2)···O8 (yellow), O3−H(O3)···O8 (dark green) and C6−HA(C6)···O8 (light green) H bonds.

Complexes **1–6** join a small family of *homometallic* Ln^{III} complexes containing triethanolamine, and its singly and doubly deprotonated forms as ligands (Table 4); Ln^{III}/tea^{3-} complexes are not known. Complexes **1**, **2** and **4** are the only dinuclear Ln^{III} complexes that possess a form of triethanolamine as the only organic ligand. Complex **6** is isomorphous with [La(NO$_3$)(teaH$_3$)$_2$](NO$_3$)$_2$ [35]. Of particular interest is the ability of teaH$_3$ to stabilize homoleptic cationic complexes with the divalent lanthanides Eu^{II} and Yb^{II} [33,34].

Table 4. Crystallographically characterized *homometallic* Ln^{II} and Ln^{III} complexes containing the $teaH_3$, $teaH_2^-$ and $teaH^{2-}$ groups as ligands.

Complex [a]	Coordination Mode [b]	Coordination Polyhedra	Ref.
$[Ln^{III}(teaH_3)_2](CF_3SO_3)_3$ (Ln = Pr, Yb, Lu)	$\eta^1{:}\eta^1{:}\eta^1{:}\eta^1$ (A)	$CSAPR^j$	[32]
$[Ln^{II}(teaH_3)_2](ClO_4)_2$ (Ln = Eu, Yb)	$\eta^1{:}\eta^1{:}\eta^1{:}\eta^1$ (A)	n.r.^{k}, $BCATAPR^l$	[33,34]
$[Ln^{III}(NO_3)_2(teaH_3)_2][NO_3]$ (Ln = La, Pr)	$\eta^1{:}\eta^1{:}\eta^1{:}\eta^1$ (A)	SPC^m	[35], this work
$[Ln^{III}(teaH_3)_2(H_2O)_2][pic]_2$ (Ln = La)^c	$\eta^1{:}\eta^1{:}\eta^1{:}\eta^1$ (A)	$BCSASPR^n$	[36]
$[Ln_6^{III}(NO_3)_6(teaH_3)_6]$ (Ln = Gd, Dy)	$\eta^1{:}\eta^1{:}\eta^2{:}\eta^2{:}\mu_3$ (E)	$SAPR^o$	[37]
$[Ln_6^{III}(CO_3)(NO_3)_2(chp)_2(teaH_2)_2(teaH_2)(H_2O)][NO_3]$ (Ln = Gd, Tb, Dy)^d	$\eta^1{:}\eta^1{:}\eta^1{:}\eta^3{:}\mu_3$ (C)^e,$\eta^1{:}\eta^1{:}\eta^1{:}\eta^2{:}\eta^2{:}\mu_3$ (E)^f	$SAPR^o$, $TCTPR^p$	[38]
$[Ln_3^{III}(OH)(teaH_2)_3(paa)_3]Cl_2$ (Ln = Dy)^g	$\eta^1{:}\eta^1{:}\eta^1{:}\eta^2{:}\mu_2$ (B)	$SAPR^o$	[39]
$[Ln_8^{III}(OH)_6(teaH)_6(teaH_2)_2(teaH_3)_2](CF_3SO_3)_4$ (Ln = Dy)^i	$\eta^1{:}\eta^1{:}\eta^1{:}\eta^3{:}\mu_3$ (G)^f,$\eta^1{:}\eta^1{:}\eta^1{:}\eta^2{:}\mu_2$ (F)^f, $\eta^1{:}\eta^1{:}\eta^1{:}\eta^1$ (A,D)^h	$SAPR^o$, $CSAPR^j$	[39]
$[Ln_2^{III}(L)(teaH_2)_6(o\text{-}van)(H_2L)(H_2O)][ClO_4]_2^i$	$\eta^1{:}\eta^1{:}\eta^1{:}\eta^2{:}\mu_2$ (B)	$TCTPR^p$	[40]
$[Ln_2^{III}(NO_3)_4(teaH_2)_2]$ (Ln = Pr, Gd, Dy)	$\eta^1{:}\eta^1{:}\eta^1{:}\eta^2{:}\mu_2$ (B)	CCU^q, $TCTPR^p$ (Ln = Gd), $CSAPR^j$ (Ln = Dy)	this work

^{a} Solvent molecules have been omitted; ^{b} See Scheme 1; ^{c} pic^- is the picrate anion; ^{d} chp^- is the anion of 6-chloro-2-hydroxypyridine; ^{e} For the $teaH_2^-$ ligands; ^{f} For the $teaH^{2-}$ ligands; ^{g} paa^- is the anion of N-(2-pyridyl)acetoacetamide; ^{h} For both the $teaH_2^-$ and $teaH_3$ ligands; ^{i} H_2L is the Schiff base N,N'-bis(3-methoxysalicylidene)-1,2-cyclohexanediamine, L^{2-} is its dianionic form and $o\text{-}van^-$ is the anion of 3-methoxysalicylaldehyde (*ortho*-vanillin); ^{j} Capped square antiprism; ^{k} For the Eu(II) complex; ^{l} Bicapped trigonal antiprism; ^{m} Sphenocorona; ^{n} Bicapped square antiprism; ^{o} Square antiprism; ^{p} Tricapped trigonal prism; ^{q} Capped cube (for the Pr_2 complex).

2.3. Magnetic Susceptibility Studies

Direct current (dc) magnetic susceptibility data (χ_M) on dried polycrystalline, analytically pure samples of **2–5** were collected in the 2.0–300 K range. The data are plotted as $\chi_M T$ vs. T products in Figure 6. The strength of the magnetic interactions between the two Ln^{III} ions in the dinuclear complexes can be easily quantified with the gadolinium analog **2**. Indeed, the Gd^{III} ions present no spin-orbit coupling at the first order. Thus, the decrease or increase of the $\chi_M T$ product when lowering the temperature for **2** reveals directly the presence of an antiferromagnetic or ferromagnetic, respectively, interaction between the Gd^{III} centers. The room temperature $\chi_M T$ value for **2** is 16.10 cm^3·K·mol^{-1}, essentially equal to the spin-only value (15.75 cm^3·K·mol^{-1}) expected for two non-interacting Gd^{III} ($^8S_{7/2}$, $S = 7/2$, $L = 0$, $g = 2$) ions. The value of the $\chi_M T$ product remains almost constant down to ~40 K and then decreases rapidly to 4.64 cm^3·K·mol^{-1} at 2.0 K, suggesting an antiferromagnetic exchange interaction. Fit of the experimental data was performed by means of the conventional analytical expression derived from the isotropic spin Hamiltonian shown in Equation (3).

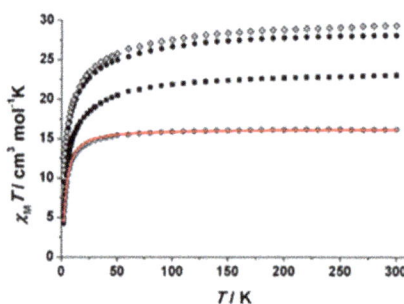

Figure 6. $\chi_M T$ vs. T plots for compounds **2** (open circles), **3** (solid squares), **4** (solid circles) and **5** (open diamonds). The solid line is the fit of the data to the theoretical model for the Gd$^{III}_2$ complex **2**; see the text for the fit parameters.

The best-fit parameters for the simultaneous simulation of susceptibility and magnetization data are $J = -0.30(1)$ and $g = 2.03$ cm^{-1} with R values of 6.1×10^{-5} ($\chi_M T$) and 1.2×10^{-4} (M). As expected for pure Ln^{III} systems, the exchange interaction is rather weak as a consequence of the shielded 4f orbitals that have small overlap with bridging ligand orbitals. This J value is typical for dinuclear complexes containing the {Gd$^{III}_2$(μ_2-OR)$_2$}$^{4+}$ core [17,18,20,25,50]. Rov and Hughbanks performed a spin density functional (SDFT) study of dinuclear Gd^{III} complexes containing the {Gd$_2$(μ_2-OR)$_2$}$^{4+}$ core [20]. The systematic study showed that symmetrically bridged complexes are antiferromagnetically coupled and asymmetrically bridged ones are ferromagnetically coupled. In the case of **2**, the {Gd$_2^{III}$(μ_2-OR)$_2$}$^{4+}$ core shows near-D_{2h} symmetry; the Gd–(μ_2-OR) bond distances are nearly equal (2.258(1) and 2.303(1) Å) and the C–(μ_2-O)–Gd angles are in the relatively narrow range 119.3(1)–130.5(1)°. Thus, the antiferromagnetic Gd^{III}···Gd^{III} exchange interaction in **2** is in accordance with the theoretical predictions [20].

$$\hat{H} = -J(\hat{S}_{Gd1} \cdot \hat{S}_{Gd1\prime}) \tag{3}$$

The room temperature $\chi_M T$ values for **3** (23.04 cm^3·K·mol^{-1}), **4** (28.07 cm^3·K·mol^{-1}) and **5** (29.32 cm^3·K·mol^{-1}) are in agreement with the expected theoretical values of 23.64, 28.34 and 28.14 cm^3·K·mol^{-1} for two non-interacting Tb^{III} ($^7F_6$, $S = 3$, $L = 3$, $g = 3/2$), Dy^{III} ($^6H_{15/2}$, $S = 5/2$, $L = 5$, $g_j = 4/3$) and Ho^{III} ($^5I_8$, $S = 2$, $L = 6$, $g_j = 5/4$) centers respectively. In all the three cases, the $\chi_M T$ product decreases slightly between 300 and ~50 K, before a more rapid decrease below ~30 K, to reach values of 4.24(**3**), 9.52(**4**) and 11.02(**5**) cm^3·K·mol^{-1} at 2.0 K. For such Ln^{III} ions with an unquenched orbital moment associated with a ligand field, the decrease of the $\chi_M T$ product as the temperature is lowered can originate from the following possible contributions [50]: (a) the thermal depopulation of the Stark sublevels; (b) the presence of magnetic anisotropy; and (c) antiferromagnetic interactions between the Ln^{III} centers. The relatively low $\chi_M T$ values at 2.0 K may suggest the contribution of a weak to moderate Ln^{III}···Ln^{III} interaction (Ln = Tb, Dy, Ho), as confirmed for the Gd$^{III}_2$ complex **2**. This assumption is further validated by the presence of a maximum in the χ_M vs. T plot at 3.3 K for the Tb$^{III}_2$ complex **3**.

Magnetization plots for **2–5** are shown in Figure 7. As a consequence of the antiferromagnetic interaction between the spin carriers, the shape of the plots is clearly sigmoid for **2** and **3**, this effect being less pronounced for **4** and almost negligible for the Ho^{III} analog **5**.

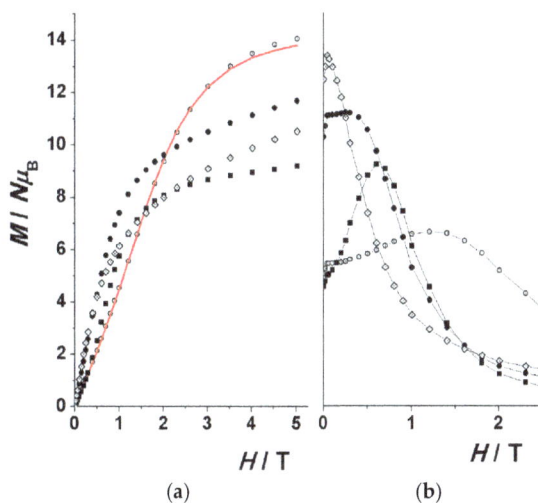

Figure 7. (a) Magnetization plots at 2.0 K for compounds **2** (open circles), **3** (solid squares), **4** (solid circles) and **5** (open diamonds). The solid line is the fit of the data to the theoretical model for the Gd$^{III}_2$ complex **2**. (b) First derivative of the magnetization plots evidencing their sigmoid shape.

In order to investigate the presence of slow relaxation of the magnetization which might originate from an SMM behavior, alternating current (ac) magnetic susceptibility measurements were performed on powdered samples of **3–5** in the temperature range 2.0–12 K with zero dc field and a 4.0 G ac field oscillating in the 10-1500 Hz range. Unfortunately, no frequency dependent out-of-phase (χ''_M) signals were detected for the complexes. This was rather surprising for **4**, because dysprosium(III) is a Kramers' ion (it has an odd number of 4f electrons), meaning that the ground state will always be bistable (one of the prerequisites for a molecule to be an SMM) irrespective of the ligand field symmetry. A review by Rinehart and Long five years ago [51] has provided a lucid account of how f-element electronic structure can, in principle, be manipulated to create new SMMS. The basic overall shape of free-ion electron density is oblate (i.e., it extends into the *xy* plane) for Tb^{III}, Dy^{III} and Ho^{III} in their ground states. Therefore, to maximize the anisotropy of an oblate ion (and thus the chances to observe SMM properties), we should place it in a crystal field for which the ligand electron density is concentrated above and below the *xy* plane; this is clearly not the case here. Given that in the absence of high symmetry (as in **4**), the ground state of Dy^{III} is a doublet along the anisotropy axis with an angular momentum quantum number $m_j = \pm 15/2$, we have determined the orientation of the ground state magnetic anisotropy axis for the Dy^{III} center of **4** using a method reported in 2013 [52], based on an electrostatic model. This method does not demand the fitting of experimental data; it only requires the knowledge of the single-crystal X-ray structure of the complex. Following this method and the program MAGELLAN (a FORTRAN program), the ground state magnetic anisotropy axis for each Dy^{III} ion (the two axis are co-parallel due to the existence of the crystallographically imposed inversion center in the molecule) is directed towards the bridging oxygen atoms O1/O1′ (Figure 8), which exhibit the shortest Dy–O bond distances (Table 2). This forces the oblate electron density of Dy^{III} to be almost parallel to the easy axis, which is a non-favorable spatial conformation to achieve slow relaxation of the magnetization [52]. No frequency dependent out-of-phase ac magnetic susceptibility signals were observed for **3–5** under external dc fields of 0.1 and 0.2 T, suggesting that the complexes are not field-induced SMMs.

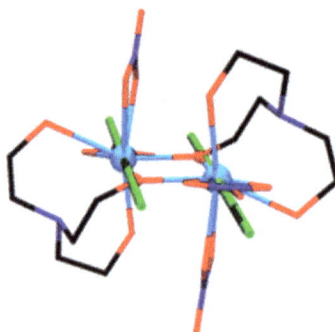

Figure 8. Ground state magnetic anisotropy axis (green bars) for the two symmetry-related Dy^{III} ions that are present in the molecule of **4**.

Since the Ln^{III} ions most often used in SMMs are Tb^{III}, Dy^{III}, Ho^{III} and Er^{III} (and rarely Yb^{III}) and since there are no reports of Pr^{III} SMMs [6], we have not studied the magnetic properties of the Pr^{III} complexes **1** and **6**.

3. Experimental Section

3.1. Materials and Physical Measurements

All manipulations were performed under aerobic conditions using reagents and solvents (Alfa Aesar, Aldrich; Karlsruhe, Germany and Tanfrichen, Germany, respectively) as received. Elemental analyses (C, H, N) were carried out by the University of Patras microanalytical service (Patras, Greece). FT-IR spectra (4000–400 cm^{-1}) were recorded using a Perkin-Elmer (supplier, Watham, MA, USA) 16 PC FT-IR spectrometer with samples prepared as KBr pellets and as nujol or hexachlorobutadiene mulls between CsI disks. Solid state, variable-temperature direct-current (dc) magnetic susceptibility data were collected on powdered samples of representative complexes using a MPMS5 Quantum Design (supplier, San Diego, CA, USA) SQUID magnetometer, operating at dc fields of 0.3 T in the 300–30 K range and 0.03 T in the 30–2.0 K range to avoid saturation effects at low temperatures. Diamagnetic corrections were applied to observed paramagnetic susceptibilities using Pascal's constants [53]. The fit of the experimental data for the dinuclear Gd^{III} complex was performed with the PHI program [54]. The quality of the fits was parameterized using the factor $R = \{(\chi_M T)_{exp} - (\chi_M T)_{calcd}\}^2 / \{(\chi_M T)_{exp}\}^2$.

3.2. Synthesis of [Pr$_2$(NO$_3$)$_4$(teaH$_2$)$_2$]·2MeOH (1·2MeOH)

To a stirred colorless solution of teaH$_3$ (66 µL, 0.50 mmol) in MeOH (20 mL) was added solid Pr(NO$_3$)$_3$·6H$_2$O (0.218 g, 0.50 mmol). The solid soon dissolved and the resulting very pale green (almost colorless) solution was stirred for a further 10 min and left undisturbed in a closed flask at room temperature. X-ray quality, pale green crystals of the product were formed over a period of 4–5 days. The crystals were collected by filtration, washed with cold MeOH (1 mL) and Et$_2$O

(3 × 1 mL), and dried in air. Typical yields were in the range 40%–45% (based on the Pr^{III} available). The complex was satisfactorily analyzed as lattice MeOH-free, i.e., as **1**. Anal calc. for $C_{12}H_{28}N_6Pr_2O_{18}$ (found values in parentheses): C 17.44 (17.63), H 3.42 (3.36), N 10.17 (9.87)%. IR bands (KBr cm^{-1}): 3356 sb, 3150 m, 2930 m, 1650 m, 1458 m, 1406 sh, 1384 s, 1094 m, 1084 m, 1064 w, 1032 m, 1004 m, 916 m, 832 w, 818 w, 738 w, 670 w, 564 w, 544 w, 526 w, 458 wb, 402 w.

3.3. Syntheses of [Gd$_2$(NO$_3$)$_4$(teaH$_2$)$_2$]·2MeOH (2·2MeOH), [Tb$_2$(NO$_3$)$_4$(teaH$_2$)$_2$]·2MeOH (3·2MeOH), [Dy$_2$(NO$_3$)$_4$(teaH$_2$)$_2$]·2MeOH (4·2MeOH) and [Ho$_2$(NO$_3$)$_4$(teaH$_2$)$_2$]·2MeOH (5·2MeOH)

These complexes were prepared and crystallized in an identical manner with **1**·2MeOH by simply replacing Pr(NO$_3$)$_3$·6H$_2$O with Gd(NO$_3$)$_3$·6H$_2$O (0.226 g, 0.50 mmol), Tb(NO$_3$)$_3$·6H$_2$O (0.227 g, 0.50 mmol), Dy(NO$_3$)$_3$·5H$_2$O (0.219 g, 0.50 mmol) and Ho(NO$_3$)$_3$·6H$_2$O (0.230 g, 0.50 mmol). The crystals of all the complexes were colorless. Typical yields were ~45% for **2**, ~55% for **3**, ~35% for **4** and ~40% for **5**. The complexes were satisfactorily analyzed as lattice MeOH-free. Anal. calc. for $C_{12}H_{28}N_6Ln_2O_{18}$ (found values in parentheses): **2** (Ln = Gd): C 16.78 (16.93), H 3.29 (3.40), N 9.79 (9.55)%; **3** (Ln = Tb): C 16.71 (16.39), H 3.28 (3.31), N 9.75 (9.50)%; **4** (Ln = Dy): C 16.58 (16.70), H 3.25 (3.37), N 9.67 (9.71)%; **5** (Ln = Ho): C 16.48 (16.31), H 3.23 (3.30), N 9.61 (9.42)%. The IR spectra of **2**, **3**, **4** and **5** are almost superimposable with the spectrum of **1** with a maximum wavenumber difference of ±4 cm^{-1}.

3.4. Synthesis of [Pr(NO$_3$)(teaH$_3$)$_2$](NO$_3$)$_2$ (6)

To a stirred colorless solution of teaH$_3$ (330 μL, 2.50 mmol) in MeOH (12 mL) was added dropwise a solution of Pr(NO$_3$)$_3$·6H$_2$O (0.435 g, 1.00 mmol) in the same solvent (5 mL). The resulting very pale green (almost colorless) solution was stirred for a further 15 min. The undisturbed solution was allowed to slowly evaporate in an open flask at room temperature. X-ray quality, pale green crystals of the product were grown over a period of 2–3 d. When precipitation was judged to be complete, the crystals were collected by filtration, washed with cold MeOH (0.5 mL) and Et$_2$O (5 × 1 mL), and dried in air. The yield was 62% (based on the Pr^{III} available). Anal. calc. for $C_{12}H_{30}N_5PrO_{15}$ (found values in parentheses): C 23.05 (22.71), H 4.85 (4.97), N 11.20 (11.27)%. Characteristic IR bands (KBr, cm^{-1}): ~3400 sb, 3230 m, 2940 m, 1643 m, 1473 m, 1010 m, 830 w.

3.5. Single-Crystal and Powder X-ray Crystallography

Suitable crystals of **1**·2MeOH, **2**·2MeOH, **4**·2MeOH and **6** had dimensions 0.12 × 0.18 × 0.36 mm, 0.10 × 0.17 × 0.20 mm, 0.10 × 0.26 × 0.44 mm and 0.10 × 0.16 × 0.39 mm, respectively. The crystals were taken from the mother liquor and immediately cooled to −113 °C (**1**·2MeOH, **4**·2MeOH, **6**) or to −43 °C (**2**·2MeOH). Diffraction data were collected on a Rigaku (Tokyo, Japan) R-AXIS SPIDER Image Plate diffractometer using graphite-monochromated Mo Kα radiation. Data collection (ω-scans) and processing (cell refinement, data reduction and Empirical absorption correction) were performed using the CrystalClear program package [55]. Important crystallographic data are listed in Table 1. The structures were solved by direct methods using SHELXS-97 [56] and refined by full-matrix least-squares techniques on F^2 with SHELXL-2014/6 [57]. Further experimental crystallographic details for **1**·2MeOH: $2\theta_{max} = 54.0°$, 262 parameters refined, $(\Delta/\sigma)_{max} = 0.002$, $(\Delta\varrho)_{max}/(\Delta\varrho)_{min} = 0.59/-0.44$ e Å^{-3}. Further experimental crystallographic details for **2**·2MeOH: $2\theta_{max}$ = 54.0°, 262 parameters refined, $(\Delta/\sigma)_{max} = 0.003$, $(\Delta\varrho)_{max}/(\Delta\varrho)_{min} = 0.48/-0.52$ e Å^{-3}. Further experimental crystallographic details for **4**·2MeOH: $2\theta_{max} = 54.0°$, 262 parameters refined, $(\Delta/\sigma)_{max} = 0.002$, $(\Delta\varrho)_{max}/(\Delta\varrho)_{min} = 0.55/-0.54$ e Å^{-3}. Further experimental crystallographic details for **6**: $2\theta_{max} = 54.0°$, 203 parameters refined, $(\Delta/\sigma)_{max} = 0.055$, $(\Delta\varrho)_{max}/(\Delta\varrho)_{min} = 0.46/-0.28$ e Å^{-3}. In the structure of **6**, the methylene groups defined by the C1 and C2 atoms present disorder at two sites with 0.8 and 0.2 occupancies. All H atoms, except those of the disordered part in **6**, were located by difference maps and were refined isotropically. All non-H atoms were refined anisotropically. Plots of the structures were drawn using the Diamond 3 program package [58]. The X-ray crystallographic data for the complexes in CIF formats have been deposited with CCDC (reference numbers CCDC 1522194, 1522195, 1522193 and 1522196 for **1**·2MeOH, **2**·2MeOH, **4**·2MeOH and **6**, respectively). They can be obtained free of charge at http://www.ccdc.cam.ac.uk//conts/retrieving.html, or from the Cambridge Crystallographic Data Centre, 12 Union Road, Cambridge CB2 1EZ, UK: Fax: +44-1223-336033; or e-mail: deposit@ccdc.cam.ac.uk. Powder X-ray diffraction patterns were collected on a Siemens D500 diffractometer (supplier, Zug, Switzerland) using Cu Kα radiation.

4. Conclusions and Perspectives

In this work, we have shown that the monoanion of triethanolamine can act as a bridging ligand forming dinuclear lanthanide(III) complexes of the general formula $[Ln_2(NO_3)_4(teaH_2)_2]$. Five members of this family (Ln = Pr, Gd, Tb, Dy, Ho) have been fully characterized. Use of excess of the ligand leads to mononuclear $[Ln(NO_3)(teaH_3)_2](NO_3)_2$ complexes with the early lanthanide(III) ions in which the ligand is neutral. Complexes **1**–**5** are the only dinuclear Ln^{III} complexes that possess a form of triethanolamine as the only organic ligand. Magnetic studies have shown that the $Gd^{III}{}_2$ complex is characterized by weak to moderate intramolecular, antiferromagnetic exchange interaction; this is most probably the case for the $Tb^{III}{}_2$, $Dy^{III}{}_2$ and $Ho^{III}{}_2$ members of the family. The dinuclear complexes with the anisotropic Ln^{III} atoms (Ln = Tb, Dy, Ho) do not exhibit SMM behavior; for the $Dy^{III}{}_2$ compound this has been rationalized by determining the metal ions' magnetic anisotropy axes the direction of which forces the oblate electron density of Dy^{III} to be almost parallel to the easy magnetization axis.

We are currently investigating the possibility to prepare Ln^{III} complexes with the triply deprotonated form of triethanolamine, i.e., tea^{3-}, as ligand; such compounds are not known to date (Table 4) and it is possible that tea^{3-} can stabilize high-nuclearity Ln^{III} compounds with interesting magnetic properties. Preliminary studies seem to confirm our expectations. As far as future perspectives are concerned, $Ln^{III}/RCO_2{}^-/teaH_2{}^-$ compounds have never been reported and it is currently not known if these complexes are isostructural with **1**–**5** or the better (compared with the nitrate ion) bridging ability of simple carboxylates can lead to products with other structural types and nuclearities.

Supplementary Materials: The following are available online at http://www.mdpi.com/2312-7481/3/1/5/s1, Figure S1: X-ray powder diffraction patterns of complexes **3**, **2** and **4**. Figure S2: The IR spectrum (KBr, cm^{-1}) of complex **3**. Figure S3: The IR spectrum (liquid between CsI disks, cm^{-1}) of the free teaH$_3$ ligand. Figure S4: Partially labeled plot of the molecule $[Pr_2(NO_3)_4(teaH_2)_2]$ that is present in the structure of **1**·2MeOH. Figure S5: Partially labeled plot of the molecule $[Gd_2(NO_3)_4(teaH_2)_2]$ that is present in the structure of **2**·2MeOH. Figure S6: The spherical capped square antiprismatic coordination geometry of Dy1 in the structure of **4**·2MeOH. Figure S7: The Johnson tricapped trigonal prismatic coordination geometry of Gd1 in the structure of **2**·2MeOH. Figure S8: The spherical-relaxed capped cubic coordination geometry of Pr1 in the structure of **1**·2MeOH. Table S1: Continuous Shape Measures (CShM) values for the potential coordination polyhedra of Dy1/Dy1' in the structure of complex $[Dy_2(NO_3)_4(teaH_2)_2]$·2MeOH (**4**·2MeOH). Table S2: Continuous Shape Measures (CShM) values for the potential coordination polyhedra of Gd1/Gd1' in the structure of complex $[Gd_2(NO_3)_4(teaH_2)_2]$·2MeOH (**2**·2MeOH). Table S3: Continuous Shape Measures (CShM) values for the potential coordination polyhedra of Pr1/Pr1' in the structure of complex $[Pr_2(NO_3)_4(teaH_2)_2]$·2MeOH (**1**·2MeOH). Table S4: Continuous Shape Measures (CShM) values for the potential coordination polyhedra of Pr1 in the structure of $[Pr(NO_3)(teaH_3)_2](NO_3)_2$ (**6**). Table S5: H bonds in the crystal structure of complex $[Dy_2(NO_3)_4(teaH_2)_2]$·2MeOH (**4**·2MeOH). Table S6: H bonds in the crystal structure of complex $[Pr(NO_3)(teaH_3)_2](NO_3)_2$ (**6**).

Acknowledgments: Albert Escuer and Julia Mayans thank the Ministeria de Economía y Competitividad, Project CTQ2015-63614-P for funding. Spyros P. Perlepes thanks the COST Action: CA15128-Molecular Spintronics (MOLSPIN) for encouraging his group activities in Patras.

Author Contributions: Ioannis Mylonas-Margaritis and Stavroula-Melina Sakellakou conducted the syntheses, crystallization and conventional characterization of the complexes; the former also contributed to the interpretation of the results. Catherine P. Raptopoulou and Vassilis Psycharis collected crystallographic data, solved the structures and performed structures refinement; the latter also studied the supramolecular features of the crystal structures and wrote the relevant part of the paper. Julia Mayans and Albert Escuer performed the magnetic measurements, interpreted the results and calculated the magnetic anisotropy axes of the Dy^{III} centers in complex **4**·2MeOH; the latter also wrote the relevant part of the paper. Spyros P. Perlepes coordinated the research, contributed to the interpretation of the results and wrote parts of the paper.

Magnetochemistry **2017**, *3*, 5

Conflicts of Interest: The authors declare no conflict of interest.

References

1. Luzon, J.; Sessoli, R. Lanthanides in molecular magnetism: So fascinating, so challenging. *Dalton Trans.* **2012**, *41*, 13556–13567. [CrossRef] [PubMed]
2. Benelli, C.; Gatteschi, D. Magnetism of Lanthanides in Molecular Materials with Transition-Metal Ions and Organic Radicals. *Chem. Rev.* **2002**, *102*, 2369–2387. [CrossRef] [PubMed]
3. Milios, C.J.; Winpenny, R.E.P. Cluster-Based Single-Molecule Magnets. *Struct. Bond.* **2015**, *164*, 1–109.
4. For a review on Mn SMMs, see: Bagai, R.; Christou, G. The Drosophila of single-molecule magnetism: [$Mn_{12}O_{12}(O_2CR)_{16}(H_2O)_4$]. *Chem. Soc. Rev.* **2009**, *38*, 1011–1026. [CrossRef] [PubMed]
5. Ishikawa, N.; Sugita, M.; Ishikawa, T.; Koshihara, S.-Y.; Kaizu, Y. Lanthanide Double-Decker Complexes Functioning as Magnets at the Single-Molecular Level. *J. Am. Chem. Soc.* **2003**, *125*, 8694–8695. [CrossRef] [PubMed]
6. For an excellent and comprehensive review, see: Woodruff, D.N.; Winpenny, R.E.P.; Layfield, R.A. Lanthanide Single-Molecule Magnets. *Chem. Rev.* **2013**, *113*, 5110–5138. [CrossRef] [PubMed]
7. Liddle, S.T.; van Slageren, J. Improving f-element single molecule magnets. *Chem. Soc. Rev.* **2015**, *44*, 6655–6669. [CrossRef] [PubMed]
8. Li, K.; Zhang, X.; Meng, X.; Shi, W.; Cheng, P.; Powell, A.K. Constraining the coordination geometries of lanthanide centers and magnetic building blocks in frameworks: A new strategy for molecular nanomagnets. *Chem. Soc. Rev.* **2016**, *45*, 2423–2439. [CrossRef] [PubMed]
9. Tang, J.; Zhang, P. Polynuclear lanthanide Single Molecule Magnets. In *Lanthanides and Actinides in Molecular Magnetism*, 1st ed.; Layfields, R.A., Murugesu, M., Eds.; Wiley-VCH: Berlin, Germany, 2015; pp. 61–88.
10. Habib, E.; Murugesu, M. Lessons learned from dinuclear lanthanide nano-magnets. *Chem. Soc. Rev.* **2013**, *42*, 3278–3288. [CrossRef] [PubMed]
11. Xue, S.; Guo, Y.-N.; Ungur, L.; Tang, J.; Chibotaru, L. Tuning the Magnetic Interactions and Relaxation Dynamics of Dy_2 Single-Molecule Magnets. *Chem. Eur. J.* **2015**, *21*, 14099–14106. [CrossRef] [PubMed]
12. Jiang, Y.; Brunet, G.; Holmberg, R.J.; Habib, F.; Karobkov, I.; Murugesu, M. Terminal solvent effects on the anisotropy barriers of Dy_2 systems. *Dalton Trans.* **2016**, *45*, 16709–16715. [CrossRef] [PubMed]
13. Xiong, J.; Ding, H.-Y.; Meng, Y.-S.; Gao, C.; Zhang, X.-J.; Meng, Z.-S.; Zhang, Y.-Q.; Shi, W.; Wang, B.-W.; Gao, S. Hydroxide-bridged five-coordinate Dy^{III} single-molecule magnet exhibiting the record thermal relaxation barrier of magnetization among lanthanide-only dimmers. *Chem. Sci.* **2016**. [CrossRef]
14. Rinehart, J.D.; Fang, M.; Evans, W.J.; Long, J.R. Strong exchange and magnetic blocking in N_2^{3-}-radical-bridged lanthanide complexes. *Nat. Chem.* **2011**, *3*, 538–542. [CrossRef] [PubMed]
15. Guo, Y.-N.; Xu, G.-F.; Wernsdorfer, W.; Ungur, L.; Guo, Y.; Tang, J.; Zhang, H.-J.; Chibotaru, L.F. Strong Axiality and Ising Exchange Interaction Suppress Zero-Field Tunneling of Magnetization of an Asymmetric Dy_2 Single-Molecule Magnet. *J. Am. Chem. Soc.* **2011**, *133*, 11948–11951. [CrossRef] [PubMed]
16. Gao, F.; Li, Y.-Y.; Liu, C.-M.; Li, Y.-Z.; Zuo, J.-L. A sandwich-type triple-decker lanthanide complex with mixed phthalocyanine and Schiff base ligands. *Dalton Trans.* **2013**, *42*, 11043–11046. [CrossRef] [PubMed]
17. Yi, X.; Bernot, K.; Cador, O.; Luzon, J.; Calvez, G.; Daiguebonne, C.; Guillou, O. Influence of ferromagnetic connection of Ising-type Dy^{III}-based single ion magnets on their magnetic slow relaxation. *Dalton Trans.* **2013**, *42*, 6728–6731. [CrossRef] [PubMed]
18. Nematirad, M.; Gee, W.J.; Langley, S.K.; Chilton, N.F.; Moubaraki, B.; Murray, K.S.; Batten, S.R. Single molecule magnetism in a μ-phenolato dinuclear motif ligated by heptadentate Schiff base ligands. *Dalton Trans.* **2012**, *41*, 13711–13715. [CrossRef] [PubMed]
19. Xu, G.-F.; Wang, Q.-L.; Gamez, P.; Ma, Y.; Clérac, R.; Tang, J.; Yan, S.-P.; Cheng, P.; Liao, D.-Z. A promising new route towards single-molecule magnets based on the oxalate ligand. *Chem. Commun.* **2010**, *46*, 1506–1508. [CrossRef] [PubMed]
20. Roy, L.E.; Hughbanks, T. Magnetic Coupling in Dinuclear Gd Complexes. *J. Am. Chem. Soc.* **2006**, *128*, 568–575. [CrossRef] [PubMed]
21. Luis, F.; Repollés, A.; Martínez-Pérez, M.J.; Aguilà, D.; Roubeau, O.; Zueco, D.; Alonso, P.J.; Evangelisti, M.; Camón, A.; Sesé, J.; et al. Molecular Prototypes for Spin-Based CNOT and SWAP Quantum Gates. *Phys. Rev. Lett.* **2011**, *107*, 117203-1–117203-5. [CrossRef] [PubMed]

22. Pedersen, K.S.; Ariciu, A.-M.; McAdams, S.; Weihe, H.; Bendix, J.; Tuna, F.; Piligkos, S. Towards Molecular 4f Single-Ion Magnet Qubits. *J. Am. Chem. Soc.* **2016**, *138*, 5801–5804. [CrossRef] [PubMed]

23. For example, see: Anastasiadis, N.C.; Mylonas-Margaritis, I.; Psycharis, V.; Raptopoulou, C.P.; Kalofolias, D.A.; Milios, C.J.; Klouras, N.; Perlepes, S.P. Dinuclear, tetrakis(acetato)-bridged lanthanide(III) complexes from the use of 2-acetylpyridine hydrazone. *Inorg. Chem. Commun.* **2015**, *51*, 99–102. [CrossRef]

24. Lin, P.-H.; Burchell, T.J.; Clérac, R.; Murugesu, M. Dinuclear Dysprosium(III) Single-Molecule Magnets with a large Anisotropic Barrier. *Angew. Chem. Int. Ed.* **2008**, *47*, 8848–8851. [CrossRef] [PubMed]

25. Anastasiadis, N.C.; Kalofolias, D.A.; Philippidis, A.; Tzani, S.; Raptopoulou, C.P.; Psycharis, V.; Milios, C.J.; Escuer, A.; Perlepes, S.P. A family of dinuclear lanthanide(III) complexes from the use of a tridentate Schiff base. *Dalton Trans.* **2015**, *44*, 10200–10209. [CrossRef] [PubMed]

26. Anastasiadis, N.C.; Granadeiro, C.M.; Klouras, N.; Cunha-Silva, L.; Raptopoulou, C.P.; Psycharis, V.; Bekiari, V.; Balula, S.S.; Escuer, A.; Perlepes, S.P. Dinuclear Lanthanide(III) Complexes by Metal-Ion-Assisted Hydration of Di-2-pyridyl Ketone Azine. *Inorg. Chem.* **2013**, *52*, 4145–4147. [CrossRef] [PubMed]

27. Nikolaou, H.; Terzis, A.; Raptopoulou, C.P.; Psycharis, V.; Bekiari, V.; Perlepes, S.P. Unique Dinuclear, Tetrakis (nitrato-O,O')-Bridged Lanthanide(III) Complexes from the Use of Pyridine-2-Amidoxime: Synthesis, Structural Studies and Spectroscopic Characterization. *J. Surf. Interfaces Mater.* **2014**, *2*, 311–318. [CrossRef]

28. Murugesu, M.; Mishra, A.; Wernsdorfer, W.; Abboud, K.A.; Christou, G. Mixed 3d/4d and 3d/4f metal clusters: Tetranuclear Fe$^{III}_2$M$^{III}_2$ (M^{III} = Ln, Y and Mn$^{IV}_2$M$^{III}_2$ (M = Yb,Y) complexes and the first Fe/4f single-molecule magnets. *Polyhedron* **2006**, *25*, 613–625. [CrossRef]

29. Schray, D.; Abbas, G.; Lan, Y.; Mereacre, V.; Sundt, A.; Dreiser, J.; Waldmann, O.; Kostakis, G.E.; Anson, C.E.; Powell, A.K. Combined Magnetic Susceptibility Measurements and ^{57}Fe Mössbauer Spectroscopy on a Ferromagnetic {Fe$^{III}_4$Dy$_4$} Ring. *Angew. Chem. Int. Ed.* **2010**, *49*, 5185–5188. [CrossRef] [PubMed]

30. Chilton, N.F.; Langley, S.K.; Moubaraki, B.; Murray, K.S. Synthesis, structural and magnetic studies of an isostructural family of mixed 3d/4f tetranuclear 'star' clusters. *Chem. Commun.* **2010**, *46*, 7787. [CrossRef] [PubMed]

31. Langley, S.K.; Wielechowski, D.P.; Moubaraki, B.; Abrahams, B.F.; Murray, K.S. Magnetic Exchange Effects in {Cr$^{III}_2$Dy$^{III}_2$} Single Molecule Magnets Containing Alcoholamine Ligands. *Aust. J. Chem.* **2014**, *67*, 1581–1587. [CrossRef]

32. Hahn, F.E.; Mohr, J. Synthese und Strukturen von Bis(triethanolamin)lanthanoid-Komplexen. *Chem. Ber.* **1990**, *123*, 481–484. [CrossRef]

33. Starynowicz, P. Synthesis and structure of bis(triethanolamine)europium(II) diperchlorate. *J. Alloy. Compd.* **2001**, *323–324*, 159–163. [CrossRef]

34. Starynowicz, P.; Gatner, K. An Ytterbium(II) Complex with Triethanolamine. *Z. Anorg. Allg. Chem.* **2003**, *629*, 722–726. [CrossRef]

35. Fowkes, A.; Harrison, W.T.A. (Nitrato-κ^{2}O, O')bis(triethanolamine-κ^{4}N, O,O',O'')lanthanum(III) dinitrate. *Acta Crystallogr. Sect. C* **2006**, *62*, m232–m233. [CrossRef] [PubMed]

36. Kumar, R.; Obrai, S.; Jassal, A.K.; Hundai, M.S. Supramolecular architectures of N, N, N', N'-tetrakis (2-hydroxyethyl)ethylenediamine and tris(2-hydroxyethyl)amine with La(III) picrate. *RSC Adv.* **2014**, *4*, 59248–59264. [CrossRef]

37. Langley, S.K.; Moubaraki, B.; Forsyth, G.M.; Gass, I.A.; Murray, K.S. Structure and magnetism of new lanthanide 6-wheel compounds utilizing triethanolamine as a stabilizing ligand. *Dalton Trans.* **2010**, *39*, 1705–1708. [CrossRef] [PubMed]

38. Langley, S.K.; Moubaraki, B.; Murray, K.S. Magnetic Properties of Hexanuclear Lanthanide(III) Clusters Incorporating a Central μ$_6$-Carbonate Ligand Derived from Atmospheric CO$_2$ Fixation. *Inorg. Chem.* **2012**, *51*, 3947–3949. [CrossRef] [PubMed]

39. Langley, S.K.; Moubaraki, B.; Murray, K.S. Trinuclear, octanuclear and decanuclear dysprosium(III) complexes: Synthesis, structural and magnetic studies. *Polyhedron* **2013**, *64*, 255–261. [CrossRef]

40. Zhang, L.; Zhang, P.; Zhao, L.; Lin, S.-Y.; Xue, S.; Tang, J.; Liu, Z. Two Locally Chiral Dysprosium Compounds with Salen-Type Ligands That Show Slow Magnetic Relaxation Behaviour. *Eur. J. Inorg. Chem.* **2013**, 1351–1357. [CrossRef]

41. Naini, A.A.; Young, V.; Verkade, J.G. New complexes of Triethanolamine (TEA): Novel Structural Features of [Y(TEA)$_2$](ClO$_4$)$_3$·3C$_5$H$_5$N and [Cd(TEA)$_2$](NO$_3$)$_2$. *Polyhedron* **1995**, *14*, 393–400. [CrossRef]

42. Tsantis, S.T.; Zagoraiou, E.; Savvidou, A.; Raptopoulou, C.P.; Psycharis, V.; Szyrwiel, L.; Holyńska, M.; Perlepes, S.P. Binding of oxime group to uranyl ion. *Dalton Trans.* **2016**, *45*, 9307–9319. [CrossRef] [PubMed]

43. Kitos, A.A.; Efthymiou, C.G.; Manos, M.J.; Tasiopoulos, A.J.; Nastopoulos, V.; Escuer, A.; Perlepes, S.P. Interesting Copper(II)-assisted transformations of 2-acetylpyridine and 2-benzoylpyridine. *Dalton Trans.* **2016**, *45*, 1063–1077. [CrossRef] [PubMed]

44. Hao, J.-M.; Yu, B.-Y.; Hecke, V.K.; Cui, G.-H. A series of d^{10} metal coordination polymers based on a flexible bis(2-methylbenzimidazole) ligand and different carboxylates: Synthesis, structures, photoluminescence and catalytic properties. *CrystEngComm* **2015**, *17*, 2279–2293. [CrossRef]

45. Cui, G.-H.; He, C.-H.; Jiao, C.-H.; Geng, J.-C.; Blatov, V.A. Two metal-organic frameworks with unique high-connected binodal network topologies: Synthesis, structures, and catalytic properties. *CrystEngComm* **2012**, *14*, 4210–4216. [CrossRef]

46. For example, see: Polyzou, C.D.; Nikolaou, H.; Papatriantafyllopoulou, C.; Psycharis, V.; Terzis, A.; Raptopoulou, C.P.; Escuer, A.; Perlepes, S.P. Employment of methyl 2-pyridyl ketone oxime in 3d/4f-metal chemistry: Dinuclear nickel(II)/lanthanide(III) species and complexes containing the metals in separate ions. *Dalton Trans.* **2012**, *41*, 13755–13764. [CrossRef] [PubMed]

47. Llunell, M.; Casanova, D.; Girera, J.; Alemany, P.; Alvarez, S. *SHAPE, Continuous Shape Measures Calculation*, version 2.0; Universitat de Barcelona: Barcelona, Spain, 2010.

48. Ruiz-Martínez, A.; Casanova, D.; Alvarez, S. Polyhedral Structures with an Odd Number of Vertices: Nine-Coordinate Metal Compounds. *Chem. Eur. J.* **2008**, *14*, 1291–1303. [CrossRef] [PubMed]

49. Kepert, D.L. *Inorganic Stereochemistry*; Springer: Berlin, Germany, 1982; pp. 179–187.

50. Long, J.; Habib, F.; Lin, P.-H.; Korobkov, I.; Enright, G.; Ungur, L.; Wernsdorfer, W.; Chibotaru, L.F.; Murugesu, M. Single-Molecule Magnet Behaviour for an Antiferromagnetically Superexchange-Coupled Dinuclear Dysprosium(III) Complex. *J. Am. Chem. Soc.* **2011**, *133*, 5319. [CrossRef] [PubMed]

51. Rinehart, J.D.; Long, J.R. Exploiting single-ion anisotropy in the design of f-element single-molecule magnets. *Chem. Sci.* **2011**, *2*, 2078–2085. [CrossRef]

52. Chilton, N.F.; Collison, D.; McInnes, E.J.I.; Winpenny, R.E.P.; Soncini, A. An electrostatic model for the determination of magnetic anisotropy in dysprosium complexes. *Nat. Commun.* **2013**, *4*, 2551–2557. [CrossRef] [PubMed]

53. Kettle, S.F.A. *Physical Inorganic Chemistry-A Coordination Chemistry Approach*; Oxford University Press: Oxford, UK, 1998; pp. 462–465.

54. Chilton, N.F.; Anderson, R.P.; Turner, L.D.; Soncini, A.; Murray, K.S. PHI: A Powerful New Program for the Analysis of Polynuclear d- and f-block Complexes. *J. Comput. Chem.* **2013**, *34*, 1164–1175. [CrossRef] [PubMed]

55. *CrystalClear*, version 1.40; Rigaku/MSC: The Woodlands, TX, USA, 2005.

56. Sheldrick, G.M. *SHELXS 97, Program for Structure Solution*; University of Göttingen: Göttingen, Germany, 1997.

57. Sheldrick, G.M. Crystal structure refinement with SHELXL. *Acta Crystallogr. Sect. C.* **2005**, *71*, 3–8.

58. *DIAMOND, Crystal and Molecular Structure Visualization*, version 3.1; Crystal Impact: Bonn, Germany.

magnetochemistry

MDPI

Article

Elaboration of Luminescent and Magnetic Hybrid Networks Based on Lanthanide Ions and Imidazolium Dicarboxylate Salts: Influence of the Synthesis Conditions

Pierre Farger, Cédric Leuvrey, Mathieu Gallart, Pierre Gilliot, Guillaume Rogez, Pierre Rabu * and Emilie Delahaye *

Institut de Physique et Chimie des Matériaux de Strasbourg, Université de Strasbourg, CNRS UMR 7504, F-67034 Strasbourg CEDEX 2, France; pierre.farger@ipcms.unistra.fr (P.F.); cedric.leuvrey@ipcms.unistra.fr (C.L.); mathieu.gallart@ipcms.unistra.fr (M.G.); pierre.gilliot@ipcms.unistra.fr (P.G.); guillaume.rogez@ipcms.unistra.fr (G.R.)
* Correspondence: pierre.rabu@ipcms.unistra.fr (P.R.); emilie.delahaye@ipcms.unistra.fr (E.D.); Tel.: +33-3-8810-7130 (P.R. & E.D.); Fax: +33-3-8810-7247 (P.R. & E.D.)

Academic Editor: Kevin Bernot
Received: 7 November 2016; Accepted: 13 December 2016; Published: 22 December 2016

Abstract: The syntheses and characterization of four new hybrid coordination networks based on lanthanide ions (Ln = Nd, Sm) and 1,3-carboxymethylimidazolium (L) salt in the presence of oxalic acid (H_2ox) are reported. The influence of the synthesis parameters, such as the nature of the lanthanide ion (Nd^{3+} or Sm^{3+}), the nature of the imidazolium source (chloride $[H_2L][Cl]$ or zwitterionic [HL] form) and the presence or not of oxalic acid (H_2ox), is discussed. In the presence of oxalic acid, the samarium salt gives only one compound $[Sm(L)(ox)(H_2O)] \cdot H_2O$, whatever the nature of the imidazolium ligand, while the neodymium salt leads to three different compounds, $[Nd(L)(ox)(H_2O)] \cdot H_2O$, $[Nd(L)(ox)_{0.5}(H_2O)_2][Cl]$ or $[Nd_2(L)_2(ox)(NO_3)(H_2O)_3][NO_3]$, depending on the imidazolium ligand. In the absence of oxalic acid, gels are obtained, except for the reaction between the neodymium salt and $[H_2L][Cl]$, which leads to $[Nd(L)(ox)(H_2O)] \cdot H_2O$. All compounds crystallized and their structures were determined by single crystal diffraction. The description of these new phases was consistently supported by ancillary techniques, such as powder X-ray diffraction, thermal analyses and UV-visible-near infrared spectroscopy. The luminescent and magnetic properties of the three pure compounds $[Sm(L)(ox)(H_2O)] \cdot H_2O$, $[Nd(L)(ox)(H_2O)] \cdot H_2O$ and $[Nd_2(L)_2(ox)(NO_3)(H_2O)_3][NO_3]$ were also studied.

Keywords: hybrid coordination networks; lanthanide ions; luminescence; magnetic properties

1. Introduction

Hybrid coordination polymers have been the subject of intense research for a few decades. Primarily investigated for their porosity and related properties, coordination polymers are promising for many applications, like gas separation and storage, catalysis or drug delivery, for example [1–5]. The versatility of the synthesis of such metal coordination polymers is now exploited to generate new functional hybrid networks with specific electronic properties (luminescence, magnetism, conductivity, etc.) [6–10].

Compared to the first row transition metals, the coordination number of 4f elements is more diverse. Even if a wide range of coordination numbers for lanthanides can make the prediction and control of the final structure of the networks difficult, it can be an advantage for the generation of (multi)functional systems. This (multi)functionality stems essentially from the intrinsic physical properties of the lanthanide ions. Especially due to luminescent properties, lanthanide-based compounds can be used in many potential applications, such as light-emitting devices, sensing, imaging agents in the biomedical area, as well as solar energy conversion [11–16]. Among the different networks based on lanthanides, those containing Tb^{3+} and Eu^{3+} ions are certainly the most studied since they exhibit a characteristic green and red emission, respectively [17,18]. Moreover, lanthanide ions present a strong magnetic anisotropy, which confers them interesting magnetic properties, such as single molecule magnet or single chain magnet behavior [19–21].

The synthesis of hybrid coordination networks is very versatile since different parameters, such as the solvent, the temperature, the pH, the nature and the concentration of the reactants, can have a great influence on the final product [1,22]. Recently, the synthesis of hybrid coordination networks has been realized in ionic liquid media, which is called ionothermal synthesis. The use of ionic liquids (belonging to the family of the imidazolium or ammonium salts, for example) allows obtaining new compounds that are only available in these kinds of conditions [23–25]. However, real control of ionothermal synthesis is still limited, especially because the role of ionic liquids, acting as a solvent, structuring agent or charge compensator, or even a combination of these three possibilities, is rather unpredictable. To circumvent this problem, we have chosen to design imidazolium salts functionalized by carboxylate functions to force the role of the imidazolium salts to that of the ligand. Such a method has already proven its efficiency to get hybrid networks [26–30].

We have recently reported the synthesis of two isostructural hybrid coordination networks based on transition metal ions (M = Co^{2+}, Zn^{2+}) [31] and a series of uranyl hybrid coordination networks [32] in the presence of the imidazolium dicarboxylate salt named 1,3-bis(carboxymethyl)imidazolium chloride or [H_2L][Cl]. To go further into the exploration and the understanding of such a system, we analyzed the behavior of the lanthanide ions, and in particular, we investigated the behavior of the Nd^{3+} and Sm^{3+} ions.

In this paper, we report the effect of the nature of the imidazolium salt (either [H_2L][Cl] or 2-(1-(carboxymethyl)-1H-imidazol-3-ium-3-yl)acetate denoted [HL]) on the structure of lanthanide-based compounds obtained by the reaction with $Nd(NO_3)_3 \cdot 6H_2O$ or $Sm(NO_3)_3 \cdot 6H_2O$ in the presence of oxalic acid (H_2ox) in a water/ethanol mixture. The effect of the presence of oxalic acid is also investigated.

2. Results and Discussion

2.1. Crystal Structure of [Ln(L)(ox)(H_2O)]·H_2O with Ln = Nd^{3+} or Sm^{3+}

The diffraction analysis reveals that the compounds [Nd(**L**)(ox)(H_2O)]·H_2O and [Sm(**L**)(ox)(H_2O)]·H_2O are isostructural (see Table 1 and Figure S1). Consequently, only the structure of [Nd(**L**)(ox)(H_2O)]·H_2O will be described below. [Nd(**L**)(ox)(H_2O)]·H_2O crystallizes in the triclinic space group *P*-1 (No. 2) with the parameters a = 8.010(3) Å, b = 9.203(3) Å, c = 9.523(2) Å, α = 79.910(20)°, β = 72.043(16)° and γ = 89.270(20)°.

Table 1. Crystal data and structure refinement for $[Sm(L)(ox)(H_2O)]\cdot H_2O$, $[Nd(L)(ox)(H_2O)]\cdot H_2O$, $[Nd_2(L)_2(ox)(NO_3)(H_2O)_3][NO_3]$ and $[Nd(L)(ox)_{0.5}(H_2O)_2][Cl]$.

Compound	$[Sm(L)(ox)(H_2O)]\cdot H_2O$	$[Nd(L)(ox)(H_2O)]\cdot H_2O$	$[Nd_2(L)_2(ox)(NO_3)(H_2O)_3][NO_3]$	$[Nd(L)(ox)_{0.5}(H_2O)_2][Cl]$
Formula	$C_9H_7N_2O_{10}Sm$	$C_9H_7N_2O_{10}Nd$	$C_{16}H_{14}N_6O_{21}Nd_2$	$C_8H_7N_2O_8Cl_1Nd$
Crystal size (mm³)	0.156 × 0.108 × 0.094	0.131 × 0.056 × 0.052	0.084 × 0.048 × 0.047	0.132 × 0.082 × 0.054
Formula weight (g·mol⁻¹)	453.52	447.41	914.81	451.68
Temperature (K)	293(2)	293(2)	293(2)	293(2)
Wavelength (Å)	0.71073	0.71073	0.71073	0.71073
Crystal system	Triclinic	Triclinic	Triclinic	Triclinic
Space group	P-1	P-1	P-1	P-1
Unit cell dimension				
a (Å)	7.9948(9)	8.010(3)	8.076(3)	7.9870(10)
b (Å)	9.2408(15)	9.203(3)	12.545(4)	8.534(3)
c (Å)	9.434(2)	9.5230(19)	15.713(3)	11.259(3)
α (°)	80.411(13)	79.91(2)	71.896(18)	71.961(17)
β (°)	71.829(11)	72.043(16)	82.14(2)	84.27(2)
γ (°)	89.793(10)	89.27(2)	75.62(3)	68.045(18)
V (Å³)	652.1(2)	656.8(3)	1462.7(8)	676.7(3)
Z	2	2	2	2
D_{calc} (g·cm⁻³)	2.310	2.262	2.077	2.154
Absorption coefficient (mm⁻¹)	4.345	3.981	3.585	4.040
F (0 0 0)	434	430	880	420
Index range	$-9<h<10$ $-11<k<12$ $-12<l<11$	$-10<h<6$ $-11<k<11$ $-12<l<11$	$-7<h<10$ $-13<k<16$ $-17<l<20$	$-10<h<10$ $-11<k<9$ $-14<l<14$
Collected reflections	6075	6785	15481	6346
Independent reflections (Rint)	2983 (0.0382)	3005 (0.0518)	6677 (0.0840)	3091 (0.1740)
Observed reflections ($I > 2\sigma(I)$)	2662	2725	3882	2334
Refinement method	Full matrix least square on F_2	Full matrix least square on F_2	Full matrix least square on F_2	Full matrix least square on F_2
Final R indices ($I > 2\sigma(I)$)	R1 = 0.0297, wR2 = 0.0632	R1 = 0.0289, wR2 = 0.0667	R1 = 0.0699, wR2 = 0.1366	R1 = 0.0802, wR2 = 0.1828
Final R indices (all data)	R1 = 0.0387, wR2 = 0.0674	R1 = 0.0352, wR2 = 0.0697	R1 = 0.1517, wR2 = 0.1665	R1 = 0.1134, wR2 = 0.2089
S	1.087	1.074	1.051	1.076
$(\Delta r)_{max, min}$ (e·Å⁻³)	1.436, −1.546	0.985, −1.643	4.237, −1.229	2.860, −3.995

In the refined model, the asymmetric unit (Figure 1) contains one Nd^{3+} cation, one fully-deprotonated L^- ligand, two half oxalate ligands, one coordinated water molecule and one non-coordinated water molecule distributed on two positions (O10A and O10B) with the occupancy rates of 0.54 and 0.46, respectively.

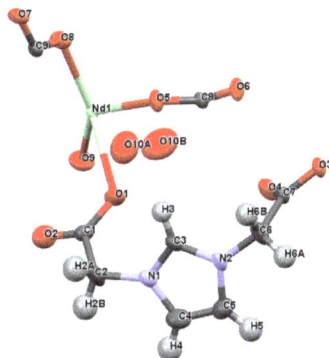

Figure 1. Ellipsoid view of the asymmetric unit of $[Nd(L)(ox)(H_2O)]\cdot H_2O$ (red: oxygen; grey: carbon; blue: nitrogen; H: hydrogen; and green: neodymium).

In the network, Nd^{3+} ions are surrounded by nine oxygens in a distorted monocapped square antiprism. Oxygen atoms belong to three different imidazolium ligands (four oxygens), two different oxalate ligands (four oxygens) and one to the coordinated water molecule. One imidazolium ligand coordinates three Nd^{3+} ions and displays a μ_3-μ_2O_3; $\kappa^2O_3O_4$; $\kappa'O_1$ coordination mode (Figure 2). One carboxylate function of the imidazolium ligand is coordinated in monodentate mode (O1). The second carboxylate function of the imidazolium ligand bridges two Nd^{3+} ions by O3 while it is coordinated to one Nd^{3+} ion with O4. The oxalate ligand coordinates two Nd^{3+} ions in bis-bidentate bridging mode as already reported [33]. These different modes of coordination give rise to dimers of lanthanide ions. These dimeric units are extended in a two-dimensional (2D) network with layers parallel to the *a0c* plan through oxalates with perpendicular orientation and di-oxo bridges (Figures 2 and 3a). The cavities of the network are filled by free water molecules almost equally distributed on the two different positions O10A and 10B (Figures 2 and 3b).

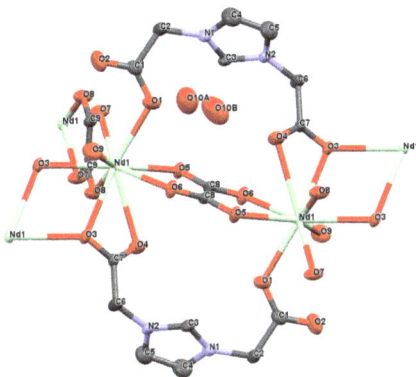

Figure 2. Selected view showing the different coordination modes in the structure of $[Nd(L)(ox)(H_2O)]\cdot H_2O$ (red: oxygen; grey: carbon; blue: nitrogen; and green: neodymium). H atoms are omitted for clarity.

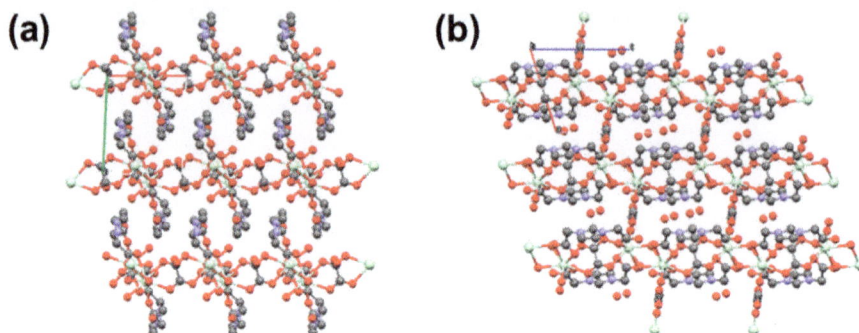

Figure 3. Selected packing view of the crystal structure of [Nd(L)(ox)(H$_2$O)]·H$_2$O showing: (a) the 2D character along the *c* axis; and (b) the cavities of the network along the *b* axis (red: oxygen; grey: carbon; blue: nitrogen; and green: neodymium). H atoms are omitted for clarity.

The Nd-O distances range from 2.413(3) Å to 2.731(3) Å (and from 2.385(4) to 2.716(3) Å for the compound [Sm(L)(ox)(H$_2$O)]·H$_2$O). These bond lengths are comparable to those observed in similar compounds [26,34]. Nd-Nd distances are equal to 4.221(1) Å (4.189(1) Å for Sm-Sm) through the O3 di-oxo bridge and 6.285(1) Å (6.234(1) Å for Sm-Sm) through the oxalate ligand.

2.2. Crystal Structure of [Nd$_2$(L)(ox)(NO$_3$)(H$_2$O)$_3$][NO$_3$]

The compound [Nd$_2$(L)(ox)(NO$_3$)(H$_2$O)$_3$][NO$_3$] is obtained as colorless crystals. It crystallizes in the triclinic space group *P*-1 (No. 2) with the parameters *a* = 8.076(3) Å, *b* = 12.545(4) Å, *c* = 15.713(3) Å, α = 71.896(18)°, β = 84.14(2)° and γ = 75.62(3)° (see Table 1 and Figure S1). The asymmetric unit contains two Nd^{3+} ions, two L$^-$ ligands, two nitrate anions (one coordinated and one non-coordinated) and three coordinated water molecules (Figure 4).

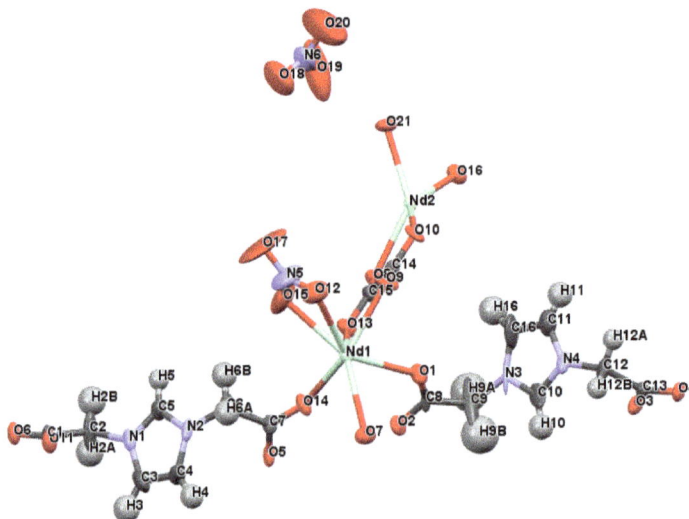

Figure 4. Ellipsoid view of the asymmetric unit of [Nd$_2$(L)(ox)(NO$_3$)(H$_2$O)$_3$][NO$_3$] (red: oxygen; grey: carbon; blue: nitrogen, H: hydrogen; and green: neodymium).

The Nd1 ion is surrounded by nine oxygen atoms belonging to two different imidazolium ligands (O1 and O11), one oxalate ligand (O13 and O9), two coordinated water molecules (O6 and O7) and one nitrate anion (O15 and O12). The Nd2 ion is also surrounded by nine oxygen atoms belonging to three different imidazolium carboxylates (O2, O3, O4, O4 and O5), one oxalate ligand (O8 and O10) and two coordinated water molecule (O16 and O21). The two Nd^{3+} ions exhibit a tricapped trigonal prism coordination polyhedron.

The Nd1 and Nd2 ions in the asymmetric unit are connected by the oxalate ligand (O10, O8, O9 and O13) in a bis-bidentate bridging mode. The carboxylate functions (O1 and O2, O14 and O5) of two different imidazolium ligands link together asymmetric units along the *a* axis.

The Nd1 ions are linked together through the carboxylate functions (O11 and O6) of two different imidazolium ligands in a bridging bidentate mode (the Nd1–Nd1 distance is 5.20 Å), whereas the Nd2 ions are interconnected by the carboxylate functions of two other ligands (the Nd2–Nd2 distance is 10.88 Å) (Figure 5). These carboxylate functions are involved in a bidentate chelate coordination mode through O3 and O4 and in a bridging bidentate coordination mode through O4 (Figure 5).

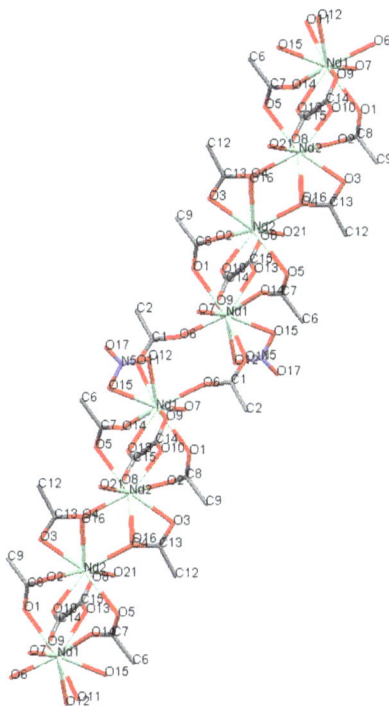

Figure 5. Selected view showing the different coordination modes for the compound [Nd$_2$(**L**)(ox)(NO$_3$)(H$_2$O)$_3$][NO$_3$] (red: oxygen; grey: carbon; blue: nitrogen; and green: neodymium). H atoms are omitted for clarity.

As for the nitrate anions, one is coordinated to Nd1 in a bidentate chelate mode, while the second is free as previously reported for the compound [Nd(**L**)$_2$(H$_2$O)$_2$][NO$_3$].3H$_2$O [26]. The bidentate chelate coordination mode of the nitrate anion is often reported in the literature [35,36]. Moreover, the imidazolium ligand is coordinated in a *trans* mode contrary to the previous structure [Ln(**L**)(ox)(H$_2$O)]·H$_2$O with Ln = Nd^{3+} or Sm^{3+} where a *cis* mode is observed. The *trans* mode gives rise to a 3D structure showing staircase linking of the Nd^{3+} ions/oxalate chains (see Figure 6). The free nitrate anions are located in the cavities of the structure. The Nd–O distances vary from

2.360(8) Å to 2.800(10) Å, and the Nd–Nd distances are equal to 4.943(2) Å and 6.413(2) Å. These distances are comparable to those observed for the previous compound [Nd(**L**)(ox)(H$_2$O)]·H$_2$O.

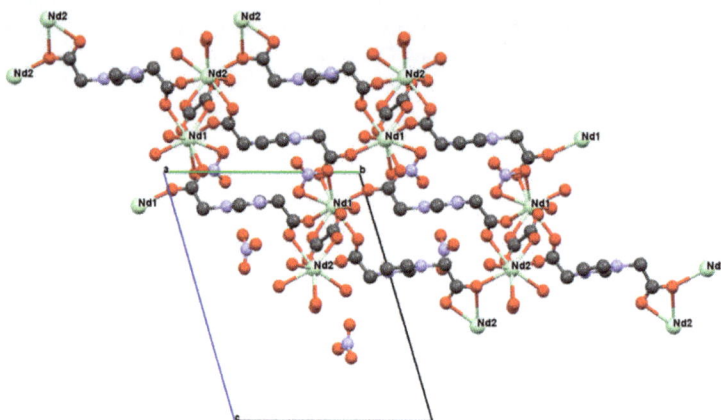

Figure 6. Selected view along the *a* axis showing different coordination modes in [Nd$_2$(**L**)(ox)(NO$_3$)(H$_2$O)$_3$] [NO$_3$] (red: oxygen; grey: carbon; blue: nitrogen; and green: neodymium). H atoms are omitted for clarity.

2.3. Crystal Structure of [Nd(L)(ox)$_{0.5}$(H$_2$O)$_2$][Cl]

The compound [Nd(**L**)(ox)$_{0.5}$(H$_2$O)$_2$][Cl] is obtained as colorless crystals. It crystallizes in the triclinic space group *P*-1 (No. 2) with the parameters *a* = 7.987(1) Å, *b* = 8.534(3) Å, *c* = 11.259(3) Å, *α* = 71.961(17)°, *β* = 84.270(20)° and *γ* = 68.045(18)° (see Table 1). The asymmetric unit contains one Nd^{3+} ion, one ligand L$^-$, one half oxalate ligand, one free chloride anion and two coordinated water molecules (Figure 7). In this structure, the Nd^{3+} ions are surrounded by nine oxygens belonging to two water molecules (O6 and O8), one oxalate ligand (O3 and O4) and three different imidazolium ligands (O1, O2, O5, O5$_{x,y,-1+z}$ and O7). The two carboxylate functions of each imidazolium ligand show different coordination modes. One carboxylate function coordinates two Nd ions through O1 and O7 in a bridging mode (O1, O7), while the second coordinates two Nd ions through O2 and O5 in a chelating bridging mode (μ_2O5; κ^2O5O2).

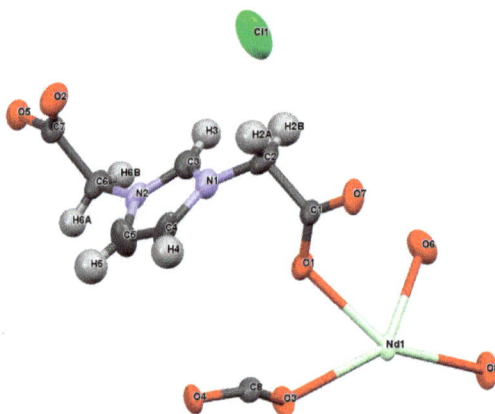

Figure 7. Asymmetric unit of [Nd(**L**)(ox)$_{0.5}$(H$_2$O)$_2$][Cl] (red: oxygen; grey: carbon; blue: nitrogen; H: hydrogen; and green: neodymium).

These different coordination modes are alternated, leading to the formation of a chain of dimeric units linked together by the oxalate ligand forming sheets parallel to the *a,b* plane. In addition, each chain is linked to another by the imidazolium ligand giving rise to a tridimensional network (Figure 8a). The chloride anion is present in the cavities of the network and is involved in hydrogen bonds with the coordinated water molecules (Figure 8b).

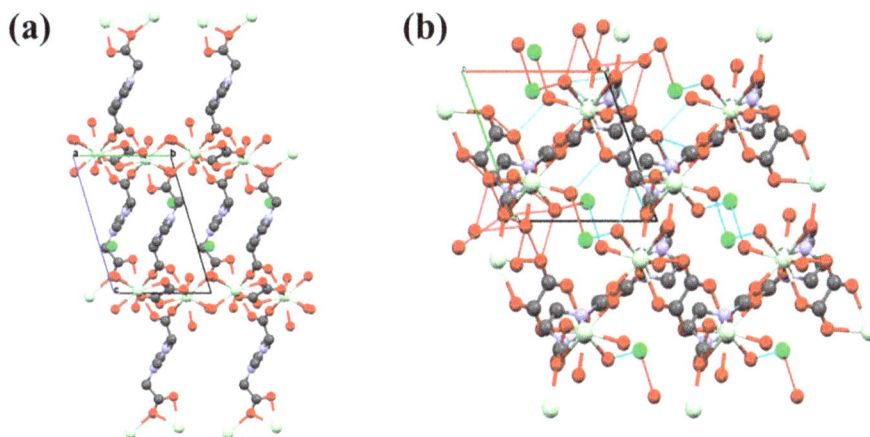

Figure 8. Selected view: (**a**) of the packing along the a axis; and (**b**) of the hydrogen bondings between the chloride anion and the networks (blue line) in $[Nd(L)(ox)_{0.5}(H_2O)_2][Cl]$ (red: oxygen; grey: carbon; blue: nitrogen; and green: neodymium). H atoms are omitted for clarity.

The Nd–O distances vary from 2.364(9) Å to 2.714(8) Å, and the Nd–Nd distances are equal to 4.987(2) Å and 6.415(2) Å through imidazolium and oxalate, respectively. These distances are similar to those observed in the previous compound $[Nd(L)(ox)(H_2O)] \cdot H_2O$. Though the crystalline structure was determined for the compound $[Nd(L)(ox)_{0.5}(H_2O)_2][Cl]$, the physical properties are not presented in the following since this compound was not obtained as a pure phase (Figure S2).

The peculiar mixed bridging-chelating coordination mode of the carboxylates encountered in both $[Nd_2(L)(ox)(NO_3)(H_2O)_3][NO_3]$ and $[Nd(L)(ox)_{0.5}(H_2O)_2][Cl]$ is reminiscent of that reported in other structures involving lanthanide ions (La^{3+} or Dy^{3+}) and linear imino diacetic acid [37]. In the latter, a "pillared" structure was observed, the ligand linking lanthanide layers; while in the present case, due to the geometry of the 1,3-carboxymethylimidazolium ligand, a staircase linking is observed between adjacent chains.

All features in the FTIR powder spectra are consistent with the single crystal structures (Figure S3) and the SEM analyses in composition mode confirm the composition of the different structures (Figure S4).

2.4. Thermal Analyses

The thermal analyses of the three compounds $[Nd_2(L)(ox)(NO_3)(H_2O)_3][NO_3]$, $[Nd(L)(ox)(H_2O)] \cdot H_2O$ and $[Sm(L)(ox)(H_2O)] \cdot H_2O$ are reported in Figure 9.

The thermal analysis of $[Sm(L)(ox)(H_2O)] \cdot H_2O$ reveals a first endothermic weight loss at 240 °C corresponding to the departure of the uncoordinated and the coordinated water molecules (calc. 7.87%; exp. 7.70%). The second weight loss observed between 300 °C and 800 °C is associated with exothermic peaks and corresponds to the decomposition of the organic species (i.e., oxalate and imidazolium ligand) and the formation of the oxide Sm_2O_3 (calc. 58.62%; exp. 57.25%). $[Nd(L)(ox)(H_2O)] \cdot H_2O$ shows a similar behavior. It shows a first exothermic weight loss at 190 °C corresponding to the loss of the water molecules (calc. 7.98%; exp. 8.98%) and a second one between 300 °C and 700 °C (calc. 59.48%; exp. 57.93%) corresponding to the decomposition of the organic species and the formation of Nd_2O_3. Concerning $[Nd_2(L)(ox)(NO_3)(H_2O)_3][NO_3]$, the first weight loss between 35 °C and 140 °C corresponds well to the departure of nitric acid (calc. 5.87%; exp. 5.43%). This loss is immediately followed by an endothermic event between 130 °C and 240 °C, which corresponds to the departure of the two water molecules and one hydroxide (calc. 6.73%; exp. 5.94%). The successive exothermic weight losses between 200 °C and 650 °C correspond to the decomposition of the organic species (i.e., imidazolium and oxalate ligands) and the nitrate concomitant with the formation of Nd_2O_3 (calc. 50.84%; exp. 52.79%).

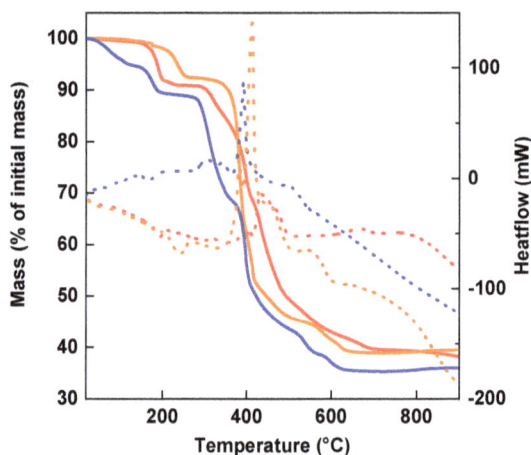

Figure 9. TGA (solid lines) and TDA (doted lines) of $[Nd_2(L)(ox)(NO_3)(H_2O)_3][NO_3]$ (blue), $[Nd(L)(ox)(H_2O)] \cdot H_2O$ (red) and $[Sm(L)(ox)(H_2O)] \cdot H_2O$ (orange).

2.5. UV-Visible-NIR Spectroscopy

The UV-visible-NIR spectra for the compounds $[Nd_2(L)(ox)(NO_3)(H_2O)_3][NO_3]$, $[Nd(L)(ox)(H_2O)] \cdot H_2O$ and $[Sm(L)(ox)(H_2O)] \cdot H_2O$ are displayed in Figure 10, while the assignments of the bands are reported on Table 2.

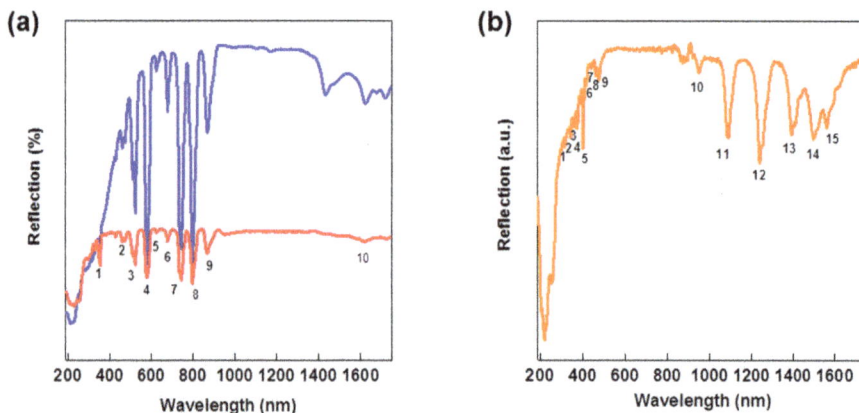

Figure 10. UV-visible-NIR spectra of the compounds: (**a**) $[Nd_2(L)(ox)(NO_3)(H_2O)_3][NO_3]$ (blue line), $[Nd(L)(ox)(H_2O)] \cdot H_2O$ (red line); and (**b**) $[Sm(L)(ox)(H_2O)] \cdot H_2O$ (orange line).

Table 2. Assignment of the bands for the compounds [Nd$_2$(**L**)(ox)(NO$_3$)(H$_2$O)$_3$][NO$_3$], [Nd(**L**)(ox)(H$_2$O)]·H$_2$O and [Sm(**L**)(ox)(H$_2$O)]·H$_2$O.

Band Number	[Nd$_2$(**L**)(ox)(NO$_3$)(H$_2$O)$_3$][NO$_3$] and [Nd(**L**)(ox)(H$_2$O)]·H$_2$O	[Sm(**L**)(ox)(H$_2$O)]·H$_2$O
1	356 nm $^4I_{9/2}{\rightarrow}^2D_{1/2}$	318 nm $^6H_{5/2}{\rightarrow}^4F_{11/2}$
2	462 nm $^4I_{9/2}{\rightarrow}^4G_{11/2}+{^2}K_{15/2}+{^2}P_{3/2}+{^2}D_{3/2}$	344 nm $^6H_{5/2}{\rightarrow}^3H_{7/2}$
3	524 nm $^4I_{9/2}{\rightarrow}^2G_{9/2}$	362 nm $^6H_{5/2}{\rightarrow}^4F_{9/2}$
4	580 nm $^4I_{9/2}{\rightarrow}^4G_{7/2}+{^2}G_{7/2}$	376 nm $^6H_{5/2}{\rightarrow}^4D_{5/2}$
5	626 nm $^4I_{9/2}{\rightarrow}^2H_{11/2}$	404 nm $^6H_{5/2}{\rightarrow}^4K_{11/2}$
6	680 nm $^4I_{9/2}{\rightarrow}^4F_{9/2}$	418 nm $^6H_{5/2}{\rightarrow}^6P_{5/2}+{^4}M_{19/2}$
7	744 nm $^4I_{9/2}{\rightarrow}^4F_{7/2},{^4}S_{3/2}$	440 nm $^6H_{5/2}{\rightarrow}^4G_{9/2}+{^4}I_{15/2}$
8	798 nm $^4I_{9/2}{\rightarrow}^4F_{5/2},{^2}H_{9/2}$	464 nm $^6H_{5/2}{\rightarrow}^4F_{5/2}+{^4}I_{13/2}$
9	870 nm $^4I_{9/2}{\rightarrow}^4F_{3/2}$	478 nm $^6H_{5/2}{\rightarrow}^4I_{11/2}+{^4}M_{15/2}$
10	1624 nm $^4I_{9/2}{\rightarrow}^4I_{15/2}$	950 nm $^6H_{5/2}{\rightarrow}^6F_{11/2}$
11	-	1088 nm $^6H_{5/2}{\rightarrow}^6F_{9/2}$
12	-	1240 nm $^6H_{5/2}{\rightarrow}^6F_{7/2}$
13	-	1390 nm $^6H_{5/2}{\rightarrow}^6F_{5/2}$
14	-	1496 nm $^6H_{5/2}{\rightarrow}^6H_{15/2}$
15	-	1562 nm $^6H_{5/2}{\rightarrow}^6F_{3/2}$

The three spectra show a common band centered at 220 nm due to the intraligand π-π^* transition of the imidazolium ligand [26,31,38].

The compounds [Nd$_2$(**L**)(ox)(NO$_3$)(H$_2$O)$_3$][NO$_3$] and [Nd(**L**)(ox)(H$_2$O)]·H$_2$O present identical bands assigned to the transitions from the ground state $^4I_{9/2}$ to the excited states $^2D_{1/2}$, $^4G_{11/2}+{^2}K_{15/2}+{^2}P_{3/2}+{^2}D_{3/2}$, $^2G_{9/2}$, $^4G_{7/2}+{^2}G_{7/2}$, $^2H_{11/2}$, $^4F_{9/2}$, $^4F_{7/2}$, $^4S_{3/2}$, $^4F_{5/2}$, $^2H_{9/2}$, $^4F_{3/2}$ and $^4I_{15/2}$ of the Nd^{3+} ion [39,40].

The compound [Sm(**L**)(ox)(H$_2$O)]·H$_2$O shows also several bands assigned to transitions from the ground state $^6H_{5/2}$ to excited states $^4F_{11/2}$, $^3H_{7/2}$, $^4F_{9/2}$, $^4D_{5/2}$ $^4K_{11/2}$, $^6P_{5/2}+{^4}M_{19/2}$, $^4F_{5/2}+{^4}I_{13/2}$, $^4I_{11/2}+{^4}M_{15/2}$, $^6F_{11/2}$, $^6F_{9/2}$, $^6F_{7/2}$, $^6F_{5/2}$, $^6H_{15/2}$ and $^6F_{3/2}$ of the Sm^{3+} ion [41,42].

2.6. Luminescent Properties

The luminescent properties of [Sm(**L**)(ox)(H$_2$O)]·H$_2$O, [Nd(**L**)(ox)(H$_2$O)]·H$_2$O and [Nd$_2$(**L**)(ox) (NO$_3$)(H$_2$O)$_3$][NO$_3$] have been investigated in the solid state at room temperature.

The luminescence (excitation and emission) spectra of [Sm(**L**)(ox)(H$_2$O)]·H$_2$O are displayed in Figure 11. The excitation spectra (monitored for λ_{em} = 404 nm) show seven peaks corresponding to the transitions from the ground state of the Sm^{3+} ion, $^6H_{5/2}$, and excited states, $^4F_{9/2}$ (364 nm), $^4D_{5/2}$ (377 nm), $^4G_{13/2}$ (391 nm), $^4K_{11/2}$ (404 nm), $^6P_{5/2}+{^4}M_{19/2}$ (417 nm), $^4G_{9/2}+{^4}I_{15/2}$ (441 nm) and $^4F_{5/2}+{^4}I_{13/2}$ (464 nm). The emission spectra (excitation at λ_{ex} = 440 nm) show the typical band for the

Sm^{3+} ion assigned to the transition from the emitting level $^4G_{5/2}$ to $^6H_{5/2}$ (572 nm), $^6H_{7/2}$ (598 nm), $^6H_{9/2}$ (644 nm), $^6H_{11/2}$ (706 nm) and $^6F_{1/2}$ (827 nm). The transition observed at 572 nm has a magnetic dipole character [40].

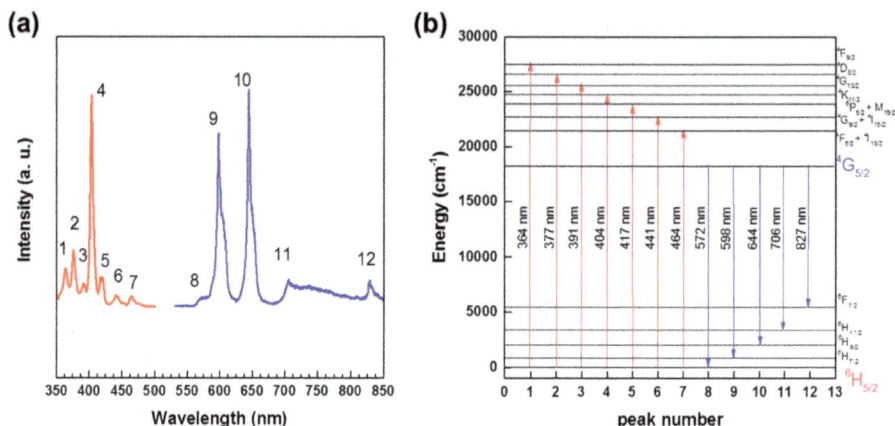

Figure 11. (**a**) Excitation spectrum (red line) and emission spectrum (blue line); and (**b**) assignment of these transitions for the compound [Sm(**L**)(ox)(H$_2$O)].

The emission spectra of [Nd$_2$(**L**)(ox)(NO$_3$)(H$_2$O)$_3$][NO$_3$] are displayed on Figure 12. The spectra show a broad band between 400 nm and 550 nm with a maximum at 433 nm, which is attributed to the luminescence of the imidazolium ligand [31]. The bands observed between 550 nm and 1100 nm are assigned to the transitions $^2G_{7/2} + {}^2G_{5/2} \rightarrow {}^4I_{9/2}$ (573 nm), $^2H_{11/2} \rightarrow {}^4I_{11/2}$ (620 nm), $^4S_{3/2} + {}^4F_{7/2} \rightarrow {}^4I_{11/2}$ (730 nm) and $^4F_{3/2} \rightarrow {}^4I_{9/2}$ (900 nm) and are characteristic of the Nd^{3+} ion [40,42].

The compound [Nd(**L**)(ox)(H$_2$O)]·H$_2$O does not display luminescence. This quenching of luminescence may be due to the presence of the uncoordinated water molecules in the interstitial sites, as previously reported [42,43].

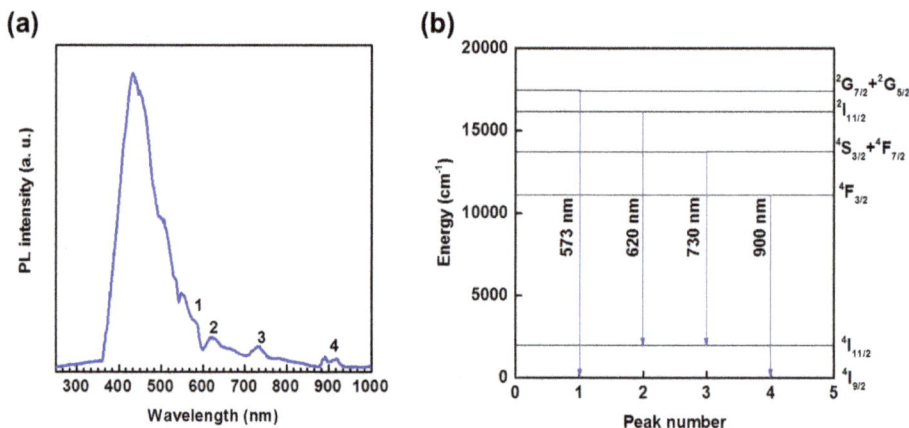

Figure 12. (**a**) Emission spectrum; and (**b**) assignment of these transitions for the compound [Nd$_2$(**L**)(ox)(NO$_3$)(H$_2$O)$_3$][NO$_3$].

2.7. Magnetic Properties

The magnetic properties of [Sm(L)(ox)(H$_2$O)]·H$_2$O and [Nd$_2$(L)(ox)(NO$_3$)(H$_2$O)$_3$][NO$_3$] were recorded under a 0.5 T DC field.

The χT product for [Sm(L)(ox)(H$_2$O)]·H$_2$O decreases linearly from 0.37 emu·K·mol^{-1} at 300 K to 0.02 emu·K·mol^{-1} at 1.8 K (see Figure 13). The value of the χT product at 300 K is the expected value for the isolated Sm^{3+} ion ($S = 5/2$, $g_J = 2/7$) [44]. At 52 K, the value of the χT product is equal to 0.09 emu·K·mol^{-1}, which is the theoretical value for the Sm^{3+} ion in its ground state, $^6H_{5/2}$ [45]. The value at 1.8 K is smaller, suggesting the presence of weak antiferromagnetic interactions between Sm^{3+} ions [33]. The Sm^{3+} ions exhibit strong spin-orbit coupling. To evaluate this coupling, we have fit the magnetic data by using a free ion approach for which the analytical expression of the susceptibility χ_M can be found [45]:

$$\chi_M = \frac{\begin{array}{c}2.143x + 7.347 + (42.92x + 1.641)\exp(-3.5x) + (283.7x - 0.6571)\exp(-8x) + \\ (620.6x - 1.94)\exp(-13.5x) + (1122x - 2.835)\exp(-20x) + (1813x - 3.556)\exp(-27.5x)\end{array}}{3 + 4\exp(-3.5x) + 5\exp(-8x) + 6\exp(-13.5x) + 7\exp(-20x) + 8\exp(-27.5x)} \tag{1}$$

where λ is the spin-orbit coupling, N the Avogadro number, β the Bohr magneton, k the Boltzmann constant and $x = \lambda/kT$.

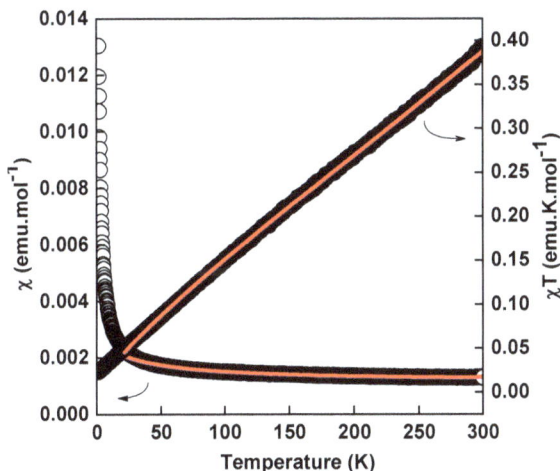

Figure 13. Plots of χ (closed circles) and χT (open circles) versus T for [Sm(L)(ox)(H$_2$O)]·H$_2$O. The red curves correspond to the fit of the data following the expressions mentioned in the text.

Fitting the χ vs. T curve with the above expression was not successful. We thus took into account a mean magnetic interaction between z neighboring Sm^{3+} ions zJ'. The expression of the susceptibility then becomes:

$$\chi = \frac{\chi_M}{1 - \chi_M \frac{2zJ'}{Ng_J^2\beta^2}} \tag{2}$$

where g_J is the Zeeman factor for Sm^{3+} ions. A good fit of the magnetic data was obtained between 300 K and 25 K with the best refined values $\lambda = 256.5(2)$ cm^{-1} and zJ' = $-4.11(3)$ cm^{-1}. These values are consistent with other values reported in the literature [33,44].

The spin-orbit coupling parameter allows then to determine the gap between the $^6H_{5/2}$ ground state and the first excited state $^6H_{7/2}$ of the Sm^{3+} ion. The gap is given by $E = 7\lambda/2 = 898(1)$ cm^{-1} [46]. This value is consistent with the value determined by the emission spectra (760 cm^{-1}) despite the free ion approximation.

The magnetic behavior of $[Nd_2(L)(ox)(NO_3)(H_2O)_3][NO_3]$ is presented in Figure 14a. The χT product decreases from 1.51 emu·K·mol^{-1} at 300 K to 0.72 emu·K·mol^{-1} at 1.8 K. The magnetic behavior of $[Nd(L)(ox)(H_2O)]\cdot H_2O$ presented in Figure 14b is similar. The χT product decreases from 1.37 emu·K·mol^{-1} at 300 K to 0.58 emu·K·mol^{-1} at 1.8 K.

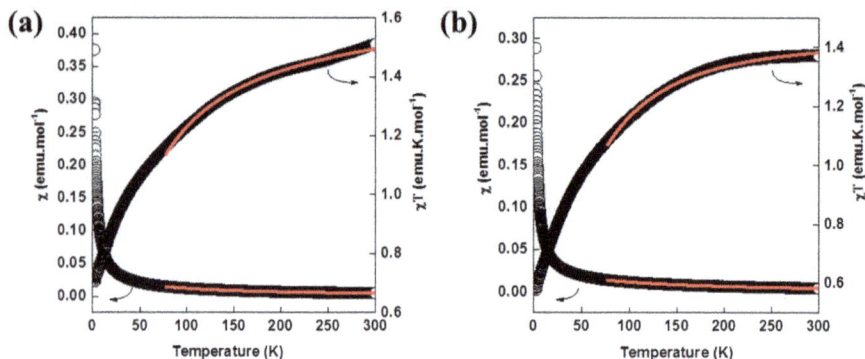

Figure 14. Plots of χ (open circles) and χT (closed circles) versus T for: (**a**) $[Nd_2(L)(ox)(NO_3)(H_2O)_3]$ $[NO_3]$; and (**b**) $[Nd(L)(ox)(H_2O)]\cdot H_2O$. The red curves represent the fit of the magnetic data.

Such behavior is commonly encountered for the isolated Nd^{3+} ion [33,47–49]. The values of the χT products at 300 K are close to that expected for the free Nd^{3+} ion (1.64 emu·K·mol^{-1} for $g_J = 8/11$). The decreasing of the χT product stems from the thermal depopulation of the low lying crystal-field states. For the Nd^{3+} ion, the first excited state is located at 2000 cm^{-1} above the ground state, and then, only the ground state is thermally populated even at 300 K. To go further, we have considered that Nd^{3+} ions may exhibit a splitting of m_J levels in an axial crystal field. The magnetic susceptibility can be described with the following expression [50]:

$$\chi = \frac{N g^2 \beta^2}{2kT} \times \frac{0.5 \exp\left(\frac{-0.25\Delta}{kT}\right) + 4.5 \exp\left(\frac{-2.25\Delta}{kT}\right) + 12.5 \exp\left(\frac{-6.25\Delta}{kT}\right) + 24.5 \exp\left(\frac{-12.25\Delta}{kT}\right) + 40.5 \exp\left(\frac{-20.25\Delta}{kT}\right)}{\exp\left(\frac{-0.25\Delta}{kT}\right) + \exp\left(\frac{-2.25\Delta}{kT}\right) + \exp\left(\frac{-6.25\Delta}{kT}\right) + \exp\left(\frac{-12.25\Delta}{kT}\right) + \exp\left(\frac{-20.25\Delta}{kT}\right)} \quad (3)$$

where Δ is the zero field splitting parameter, N the Avogadro number, β the Bohr magneton, k the Boltzmann constant and g is the Zeeman factor for Nd^{3+} ions. The fit was performed between 75 K and 300 K (red line in Figure 14a,b), and the best refined values Δ are equal to 2.79(1) cm^{-1} and 3.06(1) cm^{-1} for $[Nd_2(L)(ox)(NO_3)(H_2O)_3][NO_3]$ and $[Nd(L)(ox)(H_2O)]\cdot H_2O$, respectively. These values are slightly higher than those encountered in the literature [33,44].

2.8. Discussion

In the case of the samarium nitrate (see Figure 15), one structure $[Sm(L)(ox)(H_2O)]\cdot H_2O$ is obtained when oxalic acid is added to the reaction medium whatever the nature of the imidazolium ligand (i.e., zwitterionic [HL] or chloride salt $[H_2L][Cl]$). When the same reaction is realized without oxalic acid, the formation of a gel is observed whatever the nature of the imidazolium ligand ([HL] or $[H_2L][Cl]$).

In the case of the neodymium nitrate (see Figure 15), the situation is slightly different and more complicated since three different structures are obtained depending on the synthesis conditions. In the presence of oxalic acid, the use of imidazolium ligand in its zwitterionic form [HL] leads to the formation of the pure phase $[Nd_2(L)(ox)(NO_3)(H_2O)_3][NO_3]$, while the use of the imidazolium ligand in its chloride form $[H_2L][Cl]$ leads to a biphasic product, where the two phases have been identified as being $[Nd(L)(ox)(H_2O)]\cdot H_2O$ and $[Nd(L)(ox)_{0.5}(H_2O)_2][Cl]$. In the absence of oxalic acid, the same reaction realized with [HL] leads to a gel, while the use of $[H_2L][Cl]$ leads to crystallized $[Nd(L)(ox)(H_2O)]\cdot H_2O$, which is isostructural to $[Sm(L)(ox)(H_2O)]\cdot H_2O$. The in situ formation of the oxalate ligand is worth noticing here. Such an occurrence was previously reported and attributed to four possible main mechanisms: (i) the decomposition of the organic species [44,51,52]; (ii) the decarboxylation of the organic species followed by a reductive coupling of the carbon dioxide [53]; (iii) the oxidation of ethanol in the presence of nitrate anions [54]; and (iv) the hydrolysis followed by an oxidation and a decomposition of the organic species [55]. However, in the conditions used here in the presence of nitrates and alcohol, it is difficult to discuss these mechanisms further.

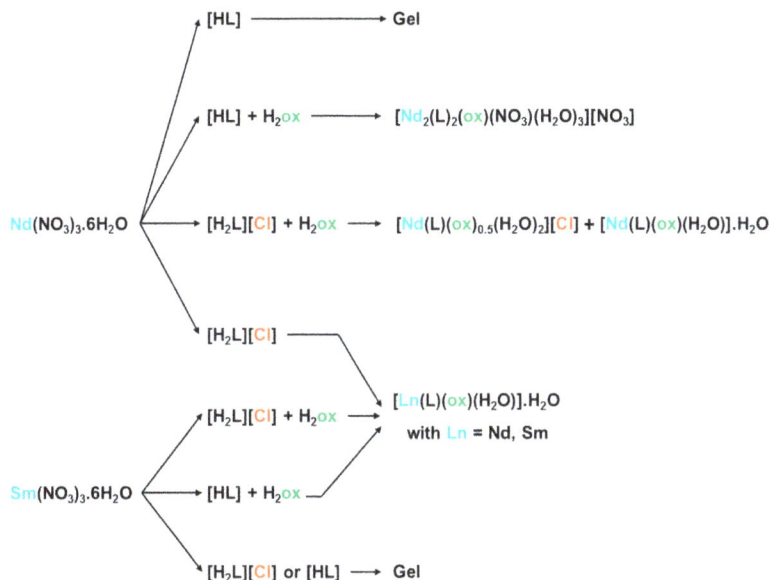

Figure 15. Recapitulative scheme indicating the compounds obtained in the water/ethanol mixture as a function of the synthesis conditions.

Nevertheless, in light of our results, it appears that the nature of the coordination networks obtained depends on three factors, which are more or less entangled: (i) the nature of the imidazolium ligand; (ii) the presence or not of oxalic acid in the parent mixture; and (iii) the nature of the lanthanide. Concerning the compounds obtained with Sm^{3+} ions, the situation is relatively simple since obtaining crystalline networks depends mainly on the presence or absence of oxalic acid. However, the situation is more complicated for the compounds obtained with Nd^{3+} ions. It is difficult to draw definitive conclusions, but it seems that the chloride anion of the imidazolium salt plays a role in the formation of the oxalate ligand and is in competition with the oxalate ligand when they are both present as starting reactants.

The modification of the obtained structures through the conditions of reaction leads for the compounds based on Nd^{3+} ions to the quenching of the luminescence properties. Indeed, in the case of $[Nd(L)(ox)(H_2O)] \cdot H_2O$, the luminescence is quenched due to the presence of free water molecules, which is not the case with $[Nd_2(L)(ox)(NO_3)(H_2O)_3][NO_3]$ possessing free nitrate anions. On the other hand, the magnetic behavior stays almost identical whatever the structure and is typical of isolated 4f ions with low antiferromagnetic interactions (through the oxalate ligand essentially) for the $[Sm(L)(ox)(H_2O)] \cdot H_2O$. The value of these antiferromagnetic interactions is in the same range than those reported in the literature [33,44]. Moreover, the detailed study of the luminescence allows corroborating the approximation of free ion used in the analysis of the magnetic data.

3. Experimental Section

3.1. Materials and Methods

1-trimethylsilylimidazole, methyl chloroacetate, glycine, paraformaldehyde, glyoxal (40%) and $Nd(NO_3)_3 \cdot 6H_2O$ were purchased from Alfa Aesar (Haverhill, MA, USA) and were used as received.

Elemental analyses for C, H and N were carried out at the Service de Microanalyses of the Institut de Chimie de Strasbourg (Strasbourg, France). The SEM images were obtained with a JEOL 67000F (Tokyo, Japan) scanning electron microscope (SEM) equipped with a field emission gun (FEG), operating at 15 kV in composition mode. FTIR spectra were collected on a Perkin Elmer Spectrum Two UATR-FTIR (Waltham, MA, USA) spectrometer. UV-Vis-NIR studies were performed on a Perkin Elmer Lambda (Waltham, MA, USA) 950 spectrometer (spectra recorded in reflection mode using a 150-mm integrating sphere with a mean resolution of 2 nm and a sampling rate of 225 $nm \cdot min^{-1}$). TGA-TDA experiments were performed on a TA instrument SDT Q600 (New Castle, DE, USA) (heating rate of 5 $°C \cdot min^{-1}$ under air stream). NMR spectra in solution were recorded using a Bruker AVANCE (Billerica, MA, USA) 300 (300 MHz) spectrometer. Photoluminescence (PL) and photoluminescence excitation (PLE) measurements were performed using a broad-spectrum Energetiq® EQ-99FC (Woburn, MA, USA) laser-driven light source (LDLS™) spectrally filtered by a monochromator. The PL signal was dispersed in a spectrometer and detected by a cooled charge coupled device (CCD) camera. Magnetic measurements were performed using a Quantum Design (Quantum Design, Inc., San Diego, CA, USA) SQUID-VSM magnetometer. The static susceptibility measurements were performed in the 1.8 K–300 K temperature range with an applied field of 0.5 T. Samples were blocked in eicosane to avoid orientation under a magnetic field. Magnetization measurements at different fields and at the given temperature confirm the absence of ferromagnetic impurities. Data were corrected for the sample holder and eicosane, and diamagnetism was estimated from Pascal constants. The powder patterns were collected on a Bruker (Billerica, MA, USA) D8 diffractometer (Cu Kα = 1.540598 Å). Details for crystal data, data collection and refinement are given in Table 1. The diffraction intensities were collected with graphite-monochromatized Mo Kα radiation (λ = 0.71073 Å). Data collection and cell refinement were carried out using a Kappa Nonius (Billerica, MA, USA) CCD diffractometer at room temperature. Intensity data were corrected for Lorentz-polarization and absorption factors. The structures were solved by direct methods using SIR92 [56] and refined against F^2 by full-matrix least-squares methods using SHLEXL-2013 [57] with the anisotropic displacement parameter for all non-hydrogen atoms. All calculations were performed by using the Crystal Structure crystallographic software package WINGX [58]. The structures were drawn using Mercury [59] or Diamond. All hydrogen atoms were located on the difference Fourier map and introduced into the calculations as the riding model with isotropic thermal parameters. Crystallographic data for the structures reported have been deposited in the Cambridge Crystallographic Data Centre with CCDC reference numbers 1515568, 1515569, 1515570, 1515571 for [Nd_2(**L**)$_2$(ox)(NO$_3$)(H_2O)$_3$][NO$_3$], [Nd(**L**)(ox)$_{0.5}$(H_2O)][Cl], [Nd(**L**)(ox)(H_2O)]·H_2O and [Sm(**L**)(ox)(H_2O)]·H_2O, respectively.

3.2. Synthesis

3.2.1. Synthesis of 1,3-Bis(carboxymethyl)imidazolium Chloride [H_2**L**][Cl]

[H_2**L**][Cl] was synthesized as previously described [31,60].

3.2.2. Synthesis of 2-(1-(Carboxymethyl)-1H-imidazol-3-ium-3-yl)acetate [H**L**]

[H**L**] was synthesized according to the modified protocols published in the literature [61,62].

Glycine (20 mmol), glyoxal (10 mmol) and paraformaldehyde (10 mmol) were dissolved in 10 mL of water. The mixture was heated at 90 °C during 7 h. The solution was concentrated, and then, the powder was obtained with the addition of ethanol (5 mL). The brown powder was recovered by filtration and dried overnight. Yield: 65%.

Elemental analysis for [HL]: $C_7H_8N_2O_4$ found (calc.) (%): C 45.20 (45.65), H 4.41 (4.34), N 14.88 (15.21). ^1H NMR (D_2O): 4.91 (s, 4), 7.43 (d, 2), 8.76 (s, 1) ppm. ^{13}C NMR (D_2O): 50.74, 122.86, 137.20, 170.73 ppm.

3.2.3. Synthesis of $[Nd_2(\mathbf{L})_2(ox)(NO_3)(H_2O)_3][NO_3]$

[HL] (0.5 mmol), $Nd(NO_3)_3 \cdot 6H_2O$ (0.5 mmol) and oxalic acid (0.25 mmol) were dissolved in a 1:1 water/ethanol mixture (1.5 mL). The mixture was sealed in a Teflon-lined stainless steel bomb (6 mL) and heated at 393 K for 72 h. After cooling to room temperature, the bomb was opened, and colorless crystals were recovered by filtration and washed with ethanol. Yield: 45%.

Elemental analysis for $[Nd_2(\mathbf{L})_2(ox)(NO_3)(H_2O)_3][NO_3] \cdot 2.9H_2O$: $C_{16}H_{25.8}N_6O_{23.9}Nd_2$ ($M = 972.68$ g/mol) Found (Calc.) (%): C 19.42 (19.74), H 2.53 (2.65), N 8.14 (8.63).

3.2.4. Synthesis of $[Nd(\mathbf{L})(ox)(H_2O)] \cdot H_2O$

$Nd(NO_3)_3 \cdot 6H_2O$ (0.5 mmol) and $[H_2\mathbf{L}][Cl]$ (0.5 mmol) were dissolved in a mixture water/ethanol (1:1 vol; 1.5 mL). The mixture was sealed in a Teflon-lined stainless bomb (6 mL) and heated at 393 K for 72 h. After cooling to room temperature, the bomb was opened, and colorless crystals were recovered by filtration and washed with ethanol (10 mL). Yield: 9.1%.

Elemental analyses: $[Nd(\mathbf{L})(ox)(H_2O)] \cdot 2.1H_2O$: $C_9H_{13.2}N_2O_{11.1}Nd_1$ ($M = 471.04$ g/mol) found (calc.) (%): C 22.52 (22.93), H 2.56 (2.80), N 6.32 (5.95).

3.2.5. Synthesis of $[Nd(\mathbf{L})(ox)_{0.5}(H_2O)][Cl]$

$Nd(NO_3)_3 \cdot 6H_2O$ (0.5 mmol), oxalic acid (0.25 mmol) and $[H_2\mathbf{L}][Cl]$ (0.5 mmol) were dissolved in a mixture water/ethanol (1:1 vol; 1.5 mL). The mixture was sealed in a Teflon-lined stainless bomb (6 mL) and heated at 393 K for 72 h. After cooling to room temperature, the bomb was opened, and colorless crystals of $[Nd(\mathbf{L})(ox)_{0.5}(H_2O)][Cl]$ and $[Nd(\mathbf{L})(ox)(H_2O)] \cdot H_2O$ were recovered by filtration and washed with ethanol (10 mL). These two compounds are colorless and without any shape difference. It was not possible to separate the two phases. No analysis was performed on the compounds obtained during this reaction.

3.2.6. Synthesis of $[Sm(\mathbf{L})(ox)(H_2O)] \cdot H_2O$

$Sm(NO_3)_3 \cdot 6H_2O$ (0.5 mmol), $[H_2\mathbf{L}][Cl]$ (0.5 mmol) and oxalic acid (0.25 mmol) were dissolved in a mixture water/ethanol (1:1 vol; 1.5 mL). The mixture was sealed in a Teflon-lined stainless bomb (6 mL) and heated at 393 K for 72 h. After cooling to room temperature, the bomb was opened, and colorless crystals were recovered by filtration and washed with ethanol (10 mL). Yield: 9.1%.

Elemental analyses: $[Sm(\mathbf{L})(ox)(H_2O)] \cdot H_2O$: $C_9H_{11}N_2O_{10}Sm_1$ ($M = 457.35$ g/mol) found (calc.) (%): C 23.46 (23.61), H 2.44 (2.40), N 5.89 (6.12).

4. Conclusions

The synthesis and the characterization of four new hybrid networks based on imidazolium dicarboxylate salts and Nd^{3+} or Sm^{3+} ions have been reported. The effect of the nature of the imidazolium salts, of the lanthanide ions, as well as of the presence of oxalic acid has been highlighted for the obtained networks. For the Sm^{3+} ions, the compound $[Sm(\mathbf{L})(ox)(H_2O)] \cdot H_2O$ is obtained only in the presence of oxalic acid whatever the nature of the imidazolium salt. For the compounds based on Nd^{3+} ions, three different compounds can be obtained according to the reaction conditions. In the presence of oxalic acid, the chloride form of the imidazolium salt leads to a biphasic product constituted of $[Nd(\mathbf{L})(ox)_{0.5}(H_2O)][Cl]$ and $[Nd(\mathbf{L})(ox)(H_2O)] \cdot H_2O$, while the zwitterionic form leads to the formation of $[Nd_2(\mathbf{L})_2(ox)(NO_3)(H_2O)_3][NO_3]$. The formation of $[Nd(\mathbf{L})(ox)(H_2O)] \cdot H_2O$ is also observed in the absence of oxalic acid with the imidazolium salt in its chloride form. The modulation of the obtained structures through the conditions of the reaction leads to the quenching of the luminescent properties for the Nd^{3+}-based compounds ions, while the magnetic behavior stays almost identical.

Supplementary Materials: The following are available online at www.mdpi.com/2312-7481/3/1/1/s1, Figure S1: Comparison of the calculated pattern from single crystals X-ray data (black line) and of the experimental powder X-ray diffraction patterns for the compound (a) $[Nd_2(L)_2(ox)(NO_3)(H_2O)_3][NO_3]$ (blue line) and (b) $[Nd(L)(ox)(H_2O)]\cdot H_2O$ (red line) and $[Sm(L)(ox)(H_2O)]\cdot H_2O$ (orange line). The green vertical lines indicate the position of the calculated diffraction lines, Figure S2: Comparison of the calculated pattern from single crystals X-ray data (black line) for $[Nd(L)(ox)_{0.5}(H_2O)][Cl]$ (black line) and $[Nd(L)(ox)(H_2O)]\cdot H_2O$ (green line) and of the experimental powder X-ray diffraction pattern (red line) of the sample coming from the reaction between $[H_2L][Cl]$, $Nd(NO_3)_3\cdot 6H_2O$ and oxalic acid, Figure S3: FTIR spectra of $[Nd_2(L)_2(ox)(NO_3)(H_2O)_3][NO_3]$ (blue line), $[Nd(L)(ox)(H_2O)]\cdot H_2O$ (orange line) and of $[H_2L][Cl]$ (black line) (a) on the range between 4000 cm^{-1} and 400 cm^{-1} and (b) enlargement of the range between 3500 cm^{-1} and 2500 cm^{-1}, Figure S4: SEM images in composition for the compounds (a) $[Nd_2(L)_2(ox)(NO_3)(H_2O)_3][NO_3]$, (b) $[Sm(L)(ox)(H_2O)]\cdot H_2O$ and (c) $[Nd(L)(ox)(H_2O)]\cdot H_2O$.

Acknowledgments: The authors thank the Centre National de la Recherche Scientifique (CNRS), the Université de Strasbourg (Idex), the Labex Nanostructures en Interaction avec leur Environnement (http://www.labex-nie.com/), the Agence Nationale de la Recherche (ANR Contract No. ANR-15-CE08-0020-01) and the Centre International de Recherche aux Frontières de la Chimie (http://www.icfrc.fr) for funding. The authors are grateful to Didier Burger (Institut de Physique et Chimie des Matériaux de Strasbourg) for technical assistance and Régis Guillot (Institut de Chimie et Matériaux Moléculaires d'Orsay) for the helpful discussions.

Author Contributions: P.F. and E.D. conceived and designed the experiments; P.F. performed the experiments; C.L. performed the SEM analysis; P.F. and E.D. performed the structural resolution, M.G. and P.G. performed the measurement of luminescence; G.R. and P.R. performed the magnetic measurements and their analysis, P.F. and E.D. wrote the paper.

Conflicts of Interest: The authors declare no conflict of interest.

References

1. Ferey, G. Hybrid porous solids: Past, present, future. *Chem. Soc. Rev.* **2008**, *37*, 191–214. [CrossRef] [PubMed]
2. Kitagawa, S.; Kitaura, R.; Noro, S.-I. Functional Porous Coordination Polymers. *Angew. Chem. Int. Ed.* **2004**, *43*, 2334–2375. [CrossRef] [PubMed]
3. Janiak, C. Engineering coordination polymers towards applications. *Dalton Trans.* **2003**, 2781–2804. [CrossRef]
4. Horcajada, P.; Serre, C.; Vallet-Regi, M.; Sebban, M.; Taulelle, F.; Ferey, G. Metal–Organic Frameworks as Efficient Materials for Drug Delivery. *Angew. Chem. Int. Ed.* **2006**, *45*, 5974. [CrossRef] [PubMed]
5. Foo, M.L.; Matsuda, R.; Kitagawa, S. Functional Hybrid Porous Coordination Polymers. *Chem. Mater.* **2014**, *26*, 310–322. [CrossRef]
6. Wang, Z.; Ananias, D.; Carné-Sánchez, A.; Brites, C.D.S.; Imaz, I.; Maspoch, D.; Rocha, J.; Carlos, L.D. Lanthanide–Organic Framework Nanothermometers Prepared by Spray-Drying. *Adv. Funct. Mater.* **2015**, *25*, 2824–2830. [CrossRef]
7. Halder, G.J.; Kepert, C.J.; Moubaraki, B.; Murray, K.S.; Cashion, J.D. Guest-Dependent Spin Crossover in a Nanoporous Molecular Framework Material. *Science* **2002**, *298*, 1762–1765. [CrossRef] [PubMed]
8. Mileo, P.G.M.; Devautour-Vinot, S.; Mouchaham, G.; Faucher, F.; Guillou, N.; Vimont, A.; Serre, C.; Maurin, G. Proton-Conducting Phenolate-Based Zr Metal–Organic Framework: A Joint Experimental–Modeling Investigation. *J. Phys. Chem. C* **2016**, *120*, 24503–24510. [CrossRef]
9. Dhara, B.; Nagarkar, S.S.; Kumar, J.; Kumar, V.; Jha, P.K.; Ghosh, S.K.; Nair, S.; Ballav, N. Increase in Electrical Conductivity of MOF to Billion-Fold upon Filling the Nanochannels with Conducting Polymer. *J. Phys. Chem. Lett.* **2016**, *7*, 2945–2950. [CrossRef] [PubMed]
10. Cirera, J. Guest effect on spin-crossover frameworks. *Rev. Inorg. Chem.* **2014**, *34*, 199–216. [CrossRef]
11. Hu, Z.; Deibert, B.J.; Li, J. Luminescent metal-organic frameworks for chemical sensing and explosive detection. *Chem. Soc. Rev.* **2014**, *43*, 5815–5840. [CrossRef] [PubMed]
12. Bunzli, J.-C.G.; Piguet, C. Taking advantage of luminescent lanthanide ions. *Chem. Soc. Rev.* **2005**, *34*, 1048–1077. [CrossRef] [PubMed]
13. Lin, R.-B.; Liu, S.-Y.; Ye, J.-W.; Li, X.-Y.; Zhang, J.-P. Photoluminescent Metal–Organic Frameworks for Gas Sensing. *Adv. Sci.* **2016**, *3*. [CrossRef] [PubMed]

14. Gándara, F.; Andrés, A.D.; Gómez-Lor, B.; Gutiérrez-Puebla, E.; Iglesias, M.; Monge, M.A.; Proserpio, D.M.; Snejko, N. A Rare-Earth MOF Series: Fascinating Structure, Efficient Light Emitters, and Promising Catalysts. *Cryst. Growth Des.* **2008**, *8*, 378–380. [CrossRef]

15. Rao, X.; Huang, Q.; Yang, X.; Cui, Y.; Yang, Y.; Wu, C.; Chen, B.; Qian, G. Color tunable and white light emitting Tb^{3+} and Eu^{3+} doped lanthanide metal-organic framework materials. *J. Mater. Chem.* **2012**, *22*, 3210–3214. [CrossRef]

16. Bünzli, J.-C.G.; Eliseeva, S.V. Lanthanide NIR luminescence for telecommunications, bioanalyses and solar energy conversion. *J. Rare Earths* **2010**, *28*, 824–842. [CrossRef]

17. Picot, A.; D'Aléo, A.; Baldeck, P.L.; Grichine, A.; Duperray, A.; Andraud, C.; Maury, O. Long-Lived Two-Photon Excited Luminescence of Water-Soluble Europium Complex: Applications in Biological Imaging Using Two-Photon Scanning Microscopy. *J. Am. Chem. Soc.* **2008**, *130*, 1532–1533. [CrossRef] [PubMed]

18. Richardson, F.S. Terbium(III) and europium(III) ions as luminescent probes and stains for biomolecular systems. *Chem. Rev.* **1982**, *82*, 541–552. [CrossRef]

19. Benelli, C.; Gatteschi, D. Magnetism of Lanthanides in Molecular Materials with Transition-Metal Ions and Organic Radicals. *Chem. Rev.* **2002**, *102*, 2369–2388. [CrossRef] [PubMed]

20. Lannes, A.; Intissar, M.; Suffren, Y.; Reber, C.; Luneau, D. Terbium(III) and Yttrium(III) Complexes with Pyridine-Substituted Nitronyl Nitroxide Radical and Different β-Diketonate Ligands. Crystal Structures and Magnetic and Luminescence Properties. *Inorg. Chem.* **2014**, *53*, 9548–9560. [CrossRef] [PubMed]

21. Bernot, K.; Bogani, L.; Caneschi, A.; Gatteschi, D.; Sessoli, R. A Family of Rare-Earth-Based Single Chain Magnets: Playing with Anisotropy. *J. Am. Chem. Soc.* **2006**, *128*, 7947–7956. [CrossRef] [PubMed]

22. Stock, N.; Biswas, S. Synthesis of Metal-Organic Frameworks (MOFs): Routes to Various MOF Topologies, Morphologies, and Composites. *Chem. Rev.* **2012**, *112*, 933–969. [CrossRef] [PubMed]

23. Reichert, W.M.; Holbrey, J.D.; Vigour, K.B.; Morgan, T.D.; Broker, G.A.; Rogers, R.D. Approaches to crystallization from ionic liquids: Complex solvents-complex results, or, a strategy for controlled formation of new supramolecular architectures? *Chem. Commun.* **2006**, 4767–4779. [CrossRef]

24. Jin, K.; Huang, X.; Pang, L.; Li, J.; Appel, A.; Wherland, S. [Cu(I)(bpp)]BF$_4$: The first extended coordination network prepared solvothermally in an ionic liquid solvent. *Chem. Commun.* **2002**, 2872–2873. [CrossRef]

25. Parnham, E.R.; Morris, R.E. Ionothermal Synthesis of Zeolites, Metal–Organic Frameworks, and Inorganic–Organic Hybrids. *Acc. Chem. Res.* **2007**, *40*, 1005–1013. [CrossRef] [PubMed]

26. Chai, X.-C.; Sun, Y.-Q.; Lei, R.; Chen, Y.-P.; Zhang, S.; Cao, Y.-N.; Zhang, H.-H. A Series of Lanthanide Frameworks with a Flexible Ligand, N,N′-Diacetic Acid Imidazolium, in Different Coordination Modes. *Cryst. Growth Des.* **2009**, *10*, 658–668. [CrossRef]

27. Abrahams, B.F.; Maynard-Casely, H.E.; Robson, R.; White, K.F. Copper(ii) coordination polymers of imdc$^-$ (H$_2$imdc$^+$ = the 1,3-bis(carboxymethyl)imidazolium cation): Unusual sheet interpenetration and an unexpected single crystal-to-single crystal transformation. *CrystEngComm* **2013**, *15*, 9729–9737. [CrossRef]

28. Zhang, X.-F.; Gao, S.; Huo, L.-H.; Zhao, H. A two-dimensional cobalt(II) coordination polymer: Poly[chloro(μ-imidazole-1,3-diyldiacetato-κ^{4}O:O′:O″:O‴)cobalt(II)]. *Acta Crystallogr. Sect. E* **2006**, *62*, m3359–m3361. [CrossRef]

29. Zhang, X.-F.; Gao, S.; Huo, L.-H.; Zhao, H. Poly[[chloromanganese(II)]-μ$_4$-imidazole-1,3-diyldiacetato]. *Acta Crystallogr. Sect. E* **2006**, *62*, m3365–m3367. [CrossRef]

30. Fei, Z.; Geldbach, T.J.; Zhao, D.; Scopelliti, R.; Dyson, P.J. A Nearly Planar Water Sheet Sandwiched between Strontium-Imidazolium Carboxylate Coordination Polymers. *Inorg. Chem.* **2005**, *44*, 5200–5202. [CrossRef] [PubMed]

31. Farger, P.; Guillot, R.; Leroux, F.; Parizel, N.; Gallart, M.; Gilliot, P.; Rogez, G.; Delahaye, E.; Rabu, P. Imidazolium Dicarboxylate Based Metal–Organic Frameworks Obtained by Solvo-Ionothermal Reaction. *Eur. J. Inorg. Chem.* **2015**, *2015*, 5342–5350. [CrossRef]

32. Martin, N.P.; Falaise, C.; Volkringer, C.; Henry, N.; Farger, P.; Falk, C.; Delahaye, E.; Rabu, P.; Loiseau, T. Hydrothermal Crystallization of Uranyl Coordination Polymers Involving an Imidazolium Dicarboxylate Ligand: Effect of pH on the Nuclearity of Uranyl-Centered Subunits. *Inorg. Chem.* **2016**, *55*, 8697–8705. [CrossRef] [PubMed]

33. Feng, X.; Ling, X.-L.; Liu, L.; Song, H.-L.; Wang, L.-Y.; Ng, S.-W.; Su, B.Y. A series of 3D lanthanide frameworks constructed from aromatic multi-carboxylate ligand: Structural diversity, luminescence and magnetic properties. *Dalton Trans.* **2013**, *42*, 10292–10303. [CrossRef] [PubMed]

34. Wang, X.-J.; Cen, Z.-M.; Ni, Q.-L.; Jiang, X.-F.; Lian, H.-C.; Gui, L.-C.; Zuo, H.-H.; Wang, Z.-Y. Synthesis Structures, and Properties of Functional 2-D Lanthanide Coordination Polymers [Ln$_2$(dpa)$_2$(C$_2$O$_4$)$_2$(H$_2$O)$_2$]n (dpa = 2,2′-(2-methylbenzimidazolium-1,3-diyl)diacetate, C$_2$O$_4$$^{2-}$ = oxalate, Ln = Nd, Eu, Gd, Tb). *Cryst. Growth Design* **2010**, *10*, 2960–2968. [CrossRef]

35. Thuery, P. Neodymium(iii) d(−)-citramalate: A chiral three-dimensional framework with water-filled channels. *CrystEngComm* **2007**, *9*, 460–462. [CrossRef]

36. Zucchi, G.; Maury, O.; Thuéry, P.; Ephritikhine, M. Structural Diversity in Neodymium Bipyrimidine Compounds with Near Infrared Luminescence: From Mono- and Binuclear Complexes to Metal-Organic Frameworks. *Inorg. Chem.* **2008**, *47*, 10398–10406. [CrossRef] [PubMed]

37. Polido Legaria, E.; Topel, S.D.; Kessler, V.G.; Seisenbaeva, G.A. Molecular insights into the selective action of a magnetically removable complexone-grafted adsorbent. *Dalton Trans.* **2015**, *44*, 1273–1282. [CrossRef] [PubMed]

38. Wang, X.-W.; Han, L.; Cai, T.-J.; Zheng, Y.-Q.; Chen, J.-Z.; Deng, Q. A Novel Chiral Doubly Folded Interpenetrating 3D Metal-Organic Framework Based on the Flexible Zwitterionic Ligand. *Cryst. Growth Des.* **2007**, *7*, 1027–1030. [CrossRef]

39. Gabrielyan, V.T.; Kaminskii, A.A.; Li, L. Absorption and luminescence spectra and energy levels of Nd^{3+} and Er^{3+} ions in LiNbO$_3$ crystals. *Phys. Status Solidi A* **1970**, *3*, K37–K42. [CrossRef]

40. Sun, L.; Qiu, Y.; Liu, T.; Zhang, J.Z.; Dang, S.; Feng, J.; Wang, Z.; Zhang, H.; Shi, L. Near Infrared and Visible Luminescence from Xerogels Covalently Grafted with Lanthanide [Sm^{3+}, Yb^{3+}, Nd^{3+}, Er^{3+}, Pr^{3+}, Ho^{3+}] β-Diketonate Derivatives Using Visible Light Excitation. *ACS Appl. Mater. Interfaces* **2013**, *5*, 9585–9593. [CrossRef] [PubMed]

41. Li, Y.-C.; Chang, Y.-H.; Lin, Y.-F.; Chang, Y.-S.; Lin, Y.-J. Synthesis and luminescent properties of Ln^{3+} (Eu^{3+}, Sm^{3+}, Dy^{3+})-doped lanthanum aluminum germanate LaAlGe$_2$O$_7$ phosphors. *J. Alloys Compd.* **2007**, *439*, 367–375. [CrossRef]

42. D'Vries, R.F.; Gomez, G.E.; Hodak, J.H.; Soler-Illia, G.J.A.A.; Ellena, J. Tuning the structure, dimensionality and luminescent properties of lanthanide metal-organic frameworks under ancillary ligand influence. *Dalton Trans.* **2016**, *45*, 646–656. [CrossRef] [PubMed]

43. Hou, G.-F.; Li, H.-X.; Li, W.-Z.; Yan, P.-F.; Su, X.-H.; Li, G.-M. Two Series of Luminescent Flexible Polycarboxylate Lanthanide Coordination Complexes with Double Layer and Rectangle Metallomacrocycle Structures. *Cryst. Growth Des.* **2013**, *13*, 3374–3380. [CrossRef]

44. Cepeda, J.; Balda, R.; Beobide, G.; Castillo, O.; Fernández, J.; Luque, A.; Pérez-Yáñez, S.; Román, P. Synthetic Control to Achieve Lanthanide(III)/Pyrimidine-4,6-dicarboxylate Compounds by Preventing Oxalate Formation: Structural, Magnetic, and Luminescent Properties. *Inorg. Chem.* **2012**, *51*, 7875–7888. [CrossRef] [PubMed]

45. Andruh, M.; Bakalbassis, E.; Kahn, O.; Trombe, J.C.; Porcher, P. Structure, spectroscopic and magnetic properties of rare earth metal(III) derivatives with the 2-formyl-4-methyl-6-(N-(2-pyridylethyl)formimidoyl)phenol ligand. *Inorg. Chem.* **1993**, *32*, 1616–1622. [CrossRef]

46. Boča, R. *Theoretical Foundations of Molecular Magnetism*; Elsevier Science: Amsterdam, The Netherlands, 1999.

47. Lhoste, J.; Perez-Campos, A.; Henry, N.; Loiseau, T.; Rabu, P.; Abraham, F. Chain-like and dinuclear coordination polymers in lanthanide (Nd, Eu) oxochloride complexes with 2,2[prime or minute]:6[prime or minute],2[prime or minute][prime or minute]-terpyridine: Synthesis, XRD structure and magnetic properties. *Dalton Trans.* **2011**, *40*, 9136–9144. [CrossRef] [PubMed]

48. Guo, L.-R.; Tang, X.-L.; Ju, Z.-H.; Zhang, K.-M.; Jiang, H.-E.; Liu, W.-S. Lanthanide metal-organic frameworks constructed by asymmetric 2-nitrobiphenyl-4,4[prime or minute]-dicarboxylate ligand: Syntheses, structures, luminescence and magnetic investigations. *CrystEngComm* **2013**, *15*, 9020–9031. [CrossRef]

49. Manna, S.C.; Zangrando, E.; Bencini, A.; Benelli, C.; Chaudhuri, N.R. Syntheses, Crystal Structures, and Magnetic Properties of [Ln^{III}$_2$(Succinate)$_3$(H$_2$O)$_2$]·0.5H$_2$O [Ln = Pr, Nd, Sm, Eu, Gd, and Dy] Polymeric Networks: Unusual Ferromagnetic Coupling in Gd Derivative. *Inorg. Chem.* **2006**, *45*, 9114–9122. [CrossRef] [PubMed]

50. Kahwa, I.A.; Selbin, J.; O'Connor, C.J.; Foise, J.W.; McPherson, G.L. Magnetic and luminescence characteristics of dinuclear complexes of lanthanides and a phenolic schiff base macrocyclic ligand. *Inorg. Chim. Acta* **1988**, *148*, 265–272. [CrossRef]

51. Abrahams, B.F.; Hudson, T.A.; Robson, R. Coordination networks incorporating the in situ generated ligands [OC(CO$_2$)$_3$]$^{4-}$ and [OCH(CO$_2$)$_2$]$^{3-}$. *J. Mol. Struct.* **2006**, *796*, 2–8. [CrossRef]

52. Knope, K.E.; Kimura, H.; Yasaka, Y.; Nakahara, M.; Andrews, M.B.; Cahill, C.L. Investigation of in Situ Oxalate Formation from 2,3-Pyrazinedicarboxylate under Hydrothermal Conditions Using Nuclear Magnetic Resonance Spectroscopy. *Inorg. Chem.* **2012**, *51*, 3883–3890. [CrossRef] [PubMed]

53. Mohapatra, S.; Vayasmudri, S.; Mostafa, G.; Maji, T.K. Lanthanide (LaIII/HoIII)-oxalate open framework materials formed by in situ ligand synthesis. *J. Mol. Struct.* **2009**, *932*, 123–128. [CrossRef]

54. Evans, O.R.; Lin, W. Synthesis of Zinc Oxalate Coordination Polymers via Unprecedented Oxidative Coupling of Methanol to Oxalic Acid. *Cryst. Growth Des.* **2001**, *1*, 9–11. [CrossRef]

55. Oliveira, C.K.; de Menezes Vicenti, J.R.; Burrow, R.A.; Alves, S., Jr.; Longo, R.L.; Malvestiti, I. Exploring the mechanism of in situ formation of oxalic acid for producing mixed fumarato-oxalato lanthanide (Eu, Tb and Gd) frameworks. *Inorg. Chem. Commun.* **2012**, *22*, 54–59. [CrossRef]

56. Altomare, A.; Cascarano, G.; Giacovazzo, C.; Guagliardi, A.; Burla, M.C.; Polidori, G.; Camalli, M. SIR92—A program for automatic solution of crystal structures by direct methods. *J. Appl. Crystallogr.* **1994**, *27*, 435. [CrossRef]

57. Sheldrick, G. A short history of SHELX. *Acta Crystallogr. Sect. A* **2008**, *64*, 112–122. [CrossRef] [PubMed]

58. Farrugia, L. WinGX suite for small-molecule single-crystal crystallography. *J. Appl. Crystallogr.* **1999**, *32*, 837–838. [CrossRef]

59. Edgington, P.R.; McCabe, P.; Macrae, C.F.; Pidcock, E.; Shields, G.P.; Taylor, R.; Towler, M.; van de Streek, J. Mercury: Visualization and analysis of crystal structures. *J. Appl. Crystallogr.* **2006**, *39*, 453–457.

60. Fei, Z.; Zhao, D.; Geldbach, T.J.; Scopelliti, R.; Dyson, P.J. Brønsted Acidic Ionic Liquids and Their Zwitterions: Synthesis, Characterization and pKa Determination. *Chem. A Eur. J.* **2004**, *10*, 4886–4893. [CrossRef] [PubMed]

61. VelíŠEk, J.; DavÍDek, T.; DavÍEk, J.; TrŠKa, P.; KvasniČKa, F.; VelcovÁ, K. New Imidazoles Formed in Nonenzymatic Browning Reactions. *J. Food Sci.* **1989**, *54*, 1544–1546. [CrossRef]

62. Kühl, O.; Palm, G. Imidazolium salts from amino acids—A new route to chiral zwitterionic carbene precursors? *Tetrahedron Asymmetry* **2010**, *21*, 393–397. [CrossRef]

MDPI

St. Alban-Anlage 66

4052 Basel, Switzerland

Tel. +41 61 683 77 34

Fax +41 61 302 89 18

http://www.mdpi.com

Magnetochemistry Editorial Office

E-mail: magnetochemistry@mdpi.com

http://www.mdpi.com/journal/magnetochemistry